Harvard Economic Studies / Volume 142

Awarded the David A. Wells Prize
for the year 1970–71 and published from
the income of the David A. Wells Fund.

The studies in this series are published under the direction of
the Department of Economics of Harvard University.
The Department does not assume
responsibility for the views expressed.

Economic Maturity
and Entrepreneurial Decline
British Iron and Steel, 1870-1913

Donald N. McCloskey

Harvard University Press   Cambridge, Massachusetts   1973

To the Memory of my Father   *Robert Green McCloskey*

1916–1969

Preface

British entrepreneurs in the late nineteenth century are alleged to have performed badly, both by comparison with their fathers, who had earned for Britain before 1870 the title of Workshop of the World, and by comparison with their American and German cousins, who were engaged after 1870 in stripping her of it. By all accounts, the British iron and steel industry from 1870 to 1913 is the worst case of relative decline. This essay is a study of entrepreneurial performance in the industry and examines the industry's market structure, the growth of its demand, its choice of technique, and its productivity relative to competitors abroad. The methods are eclectic and the pieces of evidence varied, but the conclusion to which they lead is uniform: entrepreneurs in British iron and steel, from whatever perspective they are viewed, performed well. The history of the industry in the late nineteenth century is one of economic maturity, not of entrepreneurial decline. This somewhat surprising fact is what I wish to communicate to the reader: he may wish to draw for himself the broader conclusions this case in point may suggest about the role of entrepreneurship in economic growth, the consequences of economic maturity, the uses of economics in the study of economic history, or, to descend to the particular, the sociology of Britain's relative decline in the late nineteenth century and the historical origins of Britain's economic difficulties now.

The reader should be forewarned on two points of terminology. First, I use the word "entrepreneur" throughout in the general

Preface

sense of businessman or manager rather than in the restricted
sense of a *good* businessman or manager, that is, an innovating
Schumpeterian entrepreneur. The intent is to keep the intellectual
roots of the issue in view, and these have entwined around the
word "entrepreneur." Second, I take verbal liberties with the
political geography of the United Kingdom. The United Kingdom
is the relevant concept for this study (some steel was made around
Belfast and the statistics of the industry's output and of national
income relate to Ireland as well as to Great Britain), but to avoid
distracting phrases ("The performance of the iron and steel industry was disappointing to citizens of the United Kingdom"),
"Britain" is used for the place, "British" for the corresponding
adjective, and "Englishmen" for its inhabitants.

The book began as a doctoral dissertation written under the
supervision of Alexander Gerschenkron, whose high standards of
scholarship are, I fear, insufficiently reflected in the final product.
Richard Caves and Stanley Engerman commented on the manuscript in detail, and, where I have not abandoned entirely the
many notions with which they disagreed, I have at least let the
doubts they inspired in my mind creep into the expression. Parts
of the manuscript have been subject to the critical scrutiny of
Paul David, Robert Fogel, Stanley Fischer, C. K. Harley, David
Landes, Peter Lindert, Paul McGouldrick, John Meyer, Clayne
Pope, Lars Sandberg, George Stigler, and Richard Sylla, and from
this I have benefited a great deal. Parts have been presented at
the Purdue "Cliometrics" Conference, the Harvard Conference on
the New Economic History of Britain, and the seminars in economic history at Harvard University, the University of London,
the University of Chicago, and the University of Wisconsin. I
thank their participants for the stimulus to thought they gave.

Intellectual geographers will detect the influence of both Harvard and Chicago in this work. Both institutions provided material
as well as intellectual support. Harvard awarded me a Knox Fel-
lowship to do research on the dissertation in Britain for a year.
In Britain, William Ashworth, A. H. John, and F. J. Fisher gen-

viii

erously opened their seminar in economic history to an unattached graduate student. The Library of the Iron and Steel Institute was most accommodating towards prolonged demands made on its facilities. It was in this library that my wife, Joanne, and I discovered that research assistance is the foe of domestic tranquility: but it took many months of her unstinted labor for us to make this discovery. The University of Chicago demonstrated its skepticism towards barriers to entry by permitting me to teach for two years without the union card and to work on my dissertation for a good part of that time only lightly encumbered with other responsibilities. My secretaries over the long gestation period, Marilyn Gore and Alyce Monroe, have conquered many drafts with their usual efficiency and good cheer.

I thank the *Quarterly Journal of Economics,* the *Economic History Review,* Methuen & Co., and the Princeton University Press for permission to use material from my own published work; Lars Sandberg and *Explorations in Economic History* to use material from a joint article by Sandberg and me; and the Cambridge University Press to quote extensively from Duncan Burn's *The Economic History of Steelmaking.*

Donald N. McCloskey
Chicago, Illinois
March 1972

Contents

Contents

Tables

Figures

Economic Maturity and Entrepreneurial Decline

1   The Iron and Steel Industry and the

Hypothesis of Entrepreneurial Failure

The Historiographic Career of the Hypothesis of Failure. In the last quarter of the nineteenth century, it is said, the fire of British capitalism grew cold. Some aging entrepreneurial stokers gave up their chores and retired with their earnings to country homes or public sinecures. Some few remained, but only as tenders casting a coal here or there with little thought of matching the youthful energy of the Germans or Americans. By some new law of social equilibrium, the heat of British enterprise varied inversely with its foreign competition. When American industry threw tariff walls around itself, British enterprise was damped down; when German industry invaded British markets, it was damped still further. The cold furnace stood idle, a remnant from the past, falling into disrepair. Two world wars and a long depression caused its dismantlement, and Britain took her fated place among the lesser economic powers of the earth.

The metaphor of the furnace is to the point, for the managers of Britain's iron and steel furnaces have provided many of the characters in this dismal tale of entrepreneurial failure. In his study published in 1940, *The Economic History of Steelmaking, 1867–1939,* Duncan Burn set the tone for later discussion, returning throughout the book to the "personal deficiencies," "attachment to routine," and "inadequate education" of British ironmasters.[1] Burn carefully qualified each statement of the hypothesis of entrepreneurial failure, but T. H. Burnham and G. O. Hoskins in their history of the industry written at about the same time were less circumspect: "There is, in fact, good evidence to believe that the British iron and steel industry would not have declined so fast or so far during the period reviewed had the men at the head possessed greater vision and a bolder and more energetic capacity for organization, direction and administration."[2] These two books, especially Burn's, have created a presumption of failure in iron and steel that later writers have treated as a known fact to be explained rather than as a hypothesis to be tested. In the course of a discussion of the dating and explanation of the "climacteric" in British growth in the late nineteenth century, for example, D. J. Coppock remarked that "it seems clear enough that in the iron and steel industry there was a decline in entrepreneurial talent after the 1870's."[3] Again, in his comparative study of American and British technology in the period, H. J. Habakkuk took as given the inferiority of British compared with American entrepreneurship in iron and steel, attributing the relative failure to the slower growth of the British industry: "Great generals are not made in time of peace; great entrepreneurs are not made in non-expanding industries."[4]

The alleged failure of entrepreneurs in British iron and steel has been the mainstay of recent attempts to extend the hypothesis to the economy as a whole. Derek Aldcroft, for example, drew heavily on the literature of the iron and steel industry in "The Entrepreneur and the British Economy, 1870–1914," published in 1964, as did A. L. Levine in a longer development of a similar

1. *The Economic History of Steelmaking, 1867–1939: A Study in Competition* (Cambridge: Cambridge University Press, 1940), pp. 303, 213, 11.
2. *Iron and Steel in Britain, 1870–1930* (London: Allen and Unwin, 1943), p. 271.
3. "The Climacteric of the 1890's: A Critical Note," *Manchester School,* 24 (1956), 27.
4. *American and British Technology in the Nineteenth Century* (Cambridge: Cambridge University Press, 1962), p. 212.

2

argument.[5] The most persuasive and eloquent of these extensions of the hypothesis was David Landes's contribution in 1965 to the *Cambridge Economic History of Europe*.[6] Building on an earlier piece read to a conference in 1954, "Entrepreneurship in Advanced Industrial Countries: The Anglo-German Rivalry,"[7] Landes concluded, after rejecting a number of arguments in extenuation of poor British performance, that the contrast between German and American industrial triumphs and British defeats could best be explained by emphasis on "the importance of this human factor—the success of entrepreneurial and technological creativity on one side, the failure on the other."[8] The responsibility for British failures was put squarely on bad entrepreneurship:

Thus the Britain of the late nineteenth century basked complacently in the sunset of economic hegemony . . . [N]ow it was the turn of the third generation, the children of affluence, tired of the tedium of trade and flushed with the bucolic aspirations of the country gentleman . . . The weakness of British enterprise reflected this combina-

5. Derek Aldcroft, *Economic History Review,* 2nd ser., 17 (1964), 113–134; A. L. Levine, *Industrial Retardation in Britain, 1880–1914* (London: Weidenfeld and Nicholson, 1967).

6. David S. Landes, "Technological Change and Development in Western Europe, 1750–1914," chap. V in H. J. Habakkuk and M. Postan, eds., *The Cambridge Economic History of Europe,* vol. VI: *The Industrial Revolution and After: Incomes, Population and Technological Change,* pt. I (Cambridge: Cambridge University Press, 1965). See esp. pt. IV, sec. 4, "Some Reasons Why," pp. 553–584. Landes's contribution was published separately with additional material as D. S. Landes, *The Unbound Prometheus: Technological Change and Industrial Development in Western Europe from 1750 to the Present* (Cambridge: Cambridge University Press, 1969). Subsequent reference is to *The Cambridge Economic History of Europe.*

7. Presented at a conference sponsored by the Committee on Economic Growth of the Social Science Research Council and the Harvard University Research Center in Entrepreneurial History at Cambridge, Mass. in 1954 (multilith copy). An indication of the dominance of the iron and steel industry in the case for general entrepreneurial failure is that in chap. III of this manuscript for the conference ("Some Reasons Why") fully a third of the numerous footnotes documenting the assertions in the text refer to that industry.

8. Landes, "Technological Change," p. 582.

tion of amateurism and complacency . . . [T]he British manufacturer was notorious for his indifference to style, his conservativism in the face of new techniques, his reluctance to abandon the individuality of tradition for the conformity implicit in mass production.[9]

The indictment of British entrepreneurs developed by Landes and others in the last ten years has consisted of four specific charges:

(1) They were bad salesmen, especially abroad.[10]

(2) They overinvested in the old staple export industries, such as cotton and iron, and were slow to move into the industries of the future, such as chemicals, automobiles, and electrical engineering.

(3) They underinvested in the laboratories and technical personnel required for the development and exploitation of applied science.

(4) Most important, they failed to adopt in many industries the best available techniques of production, such as ring spinning in cotton textiles, the Solvay process in chemicals, mechanical cutting in coal, and a host of new techniques in iron and steel.

Although seldom without the accompaniment of countervailing testimony, the charges have a long history, well advanced by the 1890's. It had become clear to many Englishmen by then, and to some as early as the 1870's, that Germany and America had captured the economic lead. Each year the available statistics, primarily, it should be noted, statistics of foreign trade and the output of heavy industry, provided confirming evidence that the trend of slower growth established in the 1870's was to continue. Each year, it seemed, German salesmen and chemists and American plant managers and engineers performed new feats of

9. Ibid., pp. 563–564.

10. An early work in the transformation of this allegation from journalistic to scholarly opinion was R. J. S. Hoffman's, *Great Britain and the German Trade Rivalry, 1875–1914* (Philadelphia: University of Pennsylvania Press, 1933), which used British consular reports to paint a singularly unfavorable picture of British salesmanship in Latin American, Continental, and other markets invaded by the Germans.

economic magic, while British businessmen fell into the unaccustomed role of witnesses and victims. This turn of events was anticipated, to be sure—it was clear to sober observers long before the 1890's that there was something peculiar about a small island (as its inhabitants now became fond of calling it) producing more iron, coal, and cloth than other larger countries, as it had done once, but was to do so no longer—yet it was disturbing nonetheless. In these humiliating circumstances, it was natural for Englishmen to turn to anthropomorphic and military metaphors, to emotion-laden images of youthful nations usurping the place of old Britain, and of Britain's commercial peril, struggle, invasion, and defeat. Edwin Cannan, like many other partisans in the debate over the response to be made to the threat, was scornful of this usage: "[T]he first business of the teacher of economic theory is to tear to pieces and trample upon the misleading military metaphors which have been applied by sciolists to the peaceful exchange of commodities. We hear much, for example, in these days of 'England's commercial supremacy,' and of other nations 'challenging' it, and how it is our duty to 'repel the attack,' and so on. The economist asks 'what is commercial supremacy?' and there is no answer."[11]

Notwithstanding Cannan's just rage, the metaphors stuck, to reappear in the editorial columns of newspapers, in government reports, and in retrospective writings on the period. There was, further, an inference to be drawn from them: they focused attention on the men at the top. When an army is outmaneuvered, who is to blame for its defeat but its incompetent general officers? When an economy grows old, who is to blame for its decrepitude but its aging businessmen? The play on words aroused an enduring suspicion of entrepreneurial failure in Britain relative to the United States and Germany.

The suspicion was not dispelled in all minds by dispassionate

11. "The Practical Utility of Economic Science," *Economic Journal,* 12 (1902), 470.

study of the evidence. Alfred Marshall expressed an opinion widely shared by contemporary observers when he wrote, in 1903: "Sixty years ago England had . . . leadership in most branches of industry . . . It was inevitable that she should cede much . . . to the great land which attracts alert minds of all nations to sharpen their inventive and resourceful faculties by impact on one another. It was inevitable that she should yield a little of it to that land of great industrial traditions which yoked science in the service of man with unrivaled energy. It was not inevitable that she should lose so much of it as she has done."[12] Marshall's great student, Sir John Clapham (who had himself witnessed as a young man the end of British economic dominance), chided the more intemperate of the critics of liberal economic complacency for relying on the mere scale of output as an index of success or failure, yet he too found in his study of these years case after case of bad management, particularly the slow adoption of new techniques, in textiles, in coal, in the new industries, and in iron and steel.[13] The histories of the iron and steel industry undertaken by Burnham and Hoskins and Burn in the late 1930's and early 1940's gave Clapham's judgment detailed confirmation for one industry at least, and the hypothesis of entrepreneurial failure in the late nineteenth century, reinforced by Britain's later economic difficulties, became a background premise for writings on the period.

Two developments in the late 1940's and 1950's prepared the way for the explicit statement of the hypothesis by Landes and others. A school of economic history arose that argued in Schum-

12. "Fiscal Policy of International Trade," in *Official Papers of Alfred Marshall* (London: Macmillan, 1926), p. 405. Compare his *Principles of Economics,* 8th ed. (London: Macmillan, 1920), pp. 298–300, and his *Industry and Trade,* 4th ed. (London: Macmillan, 1923), pp. 86–89. See also T. Veblen, *Imperial Germany and the Industrial Revolution* (New York: Macmillan, 1915), p. 128, and J. Hobson, *Incentives in the New Industrial Order* (New York: Seltzer, 1923), pp. 78–80, among others.
13. J. H. Clapham, *An Economic History of Modern Britain,* vol. III: *Machines and National Rivalries (1887–1914)* (Cambridge: Cambridge University Press, 1938), esp. pp. 68–72 and chap. III.

6

peterian fashion that economic growth, past and present, was heavily influenced by the quality of entrepreneurship.[14] And with the development of a statistical argument that Britain had experienced a climacteric in industrial growth in the late nineteenth century, confirming in part the opinion of contemporaries, the British entrepreneur became a case in point. The originators of the metaphor of a climacteric of the 1890's, E. H. Phelps-Brown and S. J. Handfield-Jones, intended to demonstrate that the slowing of growth was not peculiar to Britain, but was a worldwide consequence of the end of a wave of innovation gathered under the rubric of "steam and steel."[15] In the subsequent discussion, however, D. J. Coppock reinstated the view that British performance was unusually poor by offering statistical evidence that the British climacteric is more properly dated much earlier, in the 1870's.[16] Although Coppock himself did not lean heavily on the hypothesis of entrepreneurial failure, from his and related work it appeared as plain as statistics could make it that the British economy and its managers in the closing decades of the nineteenth century had indeed failed.

By the time of its full statement in the 1960's, then, the hypothesis of entrepreneurial failure had attracted a good deal of supporting argument and evidence.[17] It had attracted a good deal of

14. The center for this work was the Harvard Business School, which published the movement's journal. *Explorations in Entrepreneurial History.* Aside from Schumpeter himself, the chief figure was Arthur H. Cole. His paper, "An Approach to the Study of Entrepreneurship" (*Journal of Economic History,* 6 [Supplement, 1946], 1–15), was a manifesto for the work.

15. "The Climacteric of the 1890's: A Study of the Expanding Economy," *Oxford Economic Papers,* New Ser., 4 (1952), 266–307.

16. D. J. Coppock, "The Climacteric of the 1890's: A Critical Note," pp. 1–31.

17. Opinions in the textbooks of the last decade or so are a reasonable barometer of professional opinion on the hypothesis of failure. R. S. Sayers (*A History of Economic Change in England, 1880–1939* [London: Oxford University Press, 1967]) and W. H. B. Court (*British Economic History 1870–1914, Commentary and Documents* [Cambridge: Cambridge University Press, 1965]) were largely silent on the issue. The rest devoted

criticism as well, and one measure of its importance is the frequency with which it was taken as a null hypothesis, a foil for other interpretations of British experience in the late nineteenth century. If it was true that in the staple industries of the industrial revolution British entrepreneurs appeared to have neglected profitable techniques of production and marketing, it also could be argued that in other industries they had originated and adopted them with great speed. Charles Wilson, for example, observed that in what Sir Robert Giffen had called the "miscellaneous industries and incorporeal functions"—such as the making of soap and bicycles and the provision of financial and retailing services—British entrepreneurial performance was good.[18] Detailed narrative histories of individual industries, notably S. B. Saul's studies of mechanical engineering in the period, often reached a similar conclusion.[19]

---

a good deal of attention to it. William Ashworth (*An Economic History of England 1870–1939* [London: Methuen, 1960]) thought the suggestion "that leaders of business and technology were less ingenious and adaptable than either their fathers or their foreign contemporaries . . . a very doubtful one" (p. 241). E. J. Hobsbawm (*Industry and Empire* [London: Pantheon, 1968]) agreed, finding the several versions of the sociological explanation "all quite unconvincing" (p. 153). S. B. Saul (*The Myth of the Great Depression, 1873–1896* [London: Macmillan, 1969]) was willing to grant entrepreneurship a residual role, but a small one (see pp. 46–48). Peter Mathias (*The First Industrial Nation* [London: Methuen, 1969]) was more sympathetic to the hypothesis: "Undoubtedly, however, such failure to innovate was widespread and undoubtedly the more aggressive adoption of new techniques would have led to greater industrial investment and possibly to better records in exports" (p. 415). Sidney Pollard and D. W. Crossley (*The Wealth of Britain 1085–1966* [New York: Schocken, 1969], p. 227) expressed a similar view.

18. Charles Wilson, "Economy and Society in Late Victorian England," *Economic History Review*, 2nd ser., 18 (1965), 183–198. See also his *The History of Unilever*, vol. I (New York: Praeger, 1968; first published London: Cassell, 1954), esp. pt. I.

19. "The American Impact on British Industry, 1895–1914," *Business History*, 3 (1960), 19–38; "The Motor Industry in Britain to 1914," *Business History*, 5 (1962), 22–44; "The Export Economy 1870–1914," *Yorkshire Bulletin of Economic and Social Research*, 17 (1965), 5–18; "The Market and the Development of the Mechanical Engineering Industries in Britain, 1860–1914," *Economic History Review*, 2nd ser., 20 (1967),

When the narrative did not directly contradict the assumption of the tardy adoption of new techniques it often suggested extenuating circumstances. Two of these circumstances, the burden of old industrial equipment and the deceleration in the growth of demand, received special attention in interpretive work, particularly work by economists anxious to reconcile their distaste for the sociological cast of the hypothesis of entrepreneurial failure with their belief that new techniques had in fact been slow to come to Britain. The burden of old industrial equipment was first examined by Thorstein Veblen, who applied it to the case of the slow adoption of large coal cars on British railways. The obstacle to moving to larger coal cars was, as Marvin Frankel put Veblen's argument in an influential article on the economic burdens of the past published in 1955, the "interrelatedness" of the cars with railway equipment already in place: sidings, loading equipment, the curvature of tracks, and so forth were designed for small cars, and larger cars would therefore require massive investment in these pieces of equipment as well.[20] Charles Kindleberger favored an institutional rather than a technological version of the hypothesis of interrelatedness. The difficulty, he argued, was not so much that large investments were required to overcome the disadvantages of Britain's early start in industrialization—after all, if the older equipment proved inappropriate it could in most cases be abandoned, leaving Britain in the same position as Germany or the

---

111–130; "The Machine Tool Industry in Britain to 1914," *Business History,* 9 (1968), 22–43; "The Engineering Industry," in Derek H. Aldcroft, ed., *The Development of British Industry and Foreign Competition 1875–1914* (London: Allen and Unwin, 1968). For similar judgments on other industries, see R. A. Church, "The Effect of the American Export Invasion on the British Boot and Shoe Industry," *Journal of Economic History,* 28 (1955), 223–255; and R. E. Tyson, "The Cotton Industry," and T. C. Barker, "The Glass Industry," in Aldcroft, ed., as above.

20. M. Frankel, "Obsolescence and Technological Change in a Maturing Economy," *American Economic Review,* 45 (1955), 296–319. Frankel's article prompted a critical note by D. F. Gordon, "Obsolescence and Technological Change: Comment," *American Economic Review,* 46 (1956), 646–652 (see Frankel's "Reply" following Gordon).

9

United States—but that the benefits and costs of the investments were centered in different economic units. The small British coal cars were again the chief case in point: the railways owned the tracks and sidings, but the collieries owned the coal cars, with the result that the long-standing neglect of larger cars was rational, as in the technological version of the hypothesis, from the point of view of the individual entrepreneurs involved.[21]

The second extenuating circumstance that found favor with economists wary of sociological theories of the failure to adopt new techniques was the slow growth of demand. Slower growth in Britain than in the industrializing countries meant that it was rational to keep an older capital stock: a slowly growing capital stock, like a slowly growing human population, has a higher average age and includes therefore less up-to-date components. This explanation of the allegedly antique character of British equipment was a popular alternative to entrepreneurial failure among contemporaries, especially for explaining technological lags in the steel industry, where it seemed most likely to apply, and was adopted by subsequent doubters of the entrepreneurial hypothesis such as Ingvar Svennilson and H. J. Habakkuk.[22] Like the interrelatedness argument, the age-of-capital argument received its formal theoretical baptism in the 1950's, well after it had been

21. C. P. Kindleberger, *Economic Growth in France and Britain, 1851–1950* (Cambridge: Harvard University Press, 1964), chaps. 6 and 7; see also his "Obsolescence and Technical Change," *Oxford University Institute of Statistics Bulletin,* 23 (1961), 281–297. Incidentally, the coal cars are on the face of it a doubtful example on which to base the argument of interrelatedness in either of its forms, for they are still small, twenty years after nationalization of the coal mines and the railways and eighty years after their alleged economic inferiority first emerged.

22. One contemporary proponent of the argument was S. J. Chapman, speaking of steel in 1904, and quoted with approval in H. J. Habakkuk, *American and British Technology in the 19th Century* (Cambridge: Cambridge University Press, 1962), p. 208: "The up-to-date character of many American works is as much an effect as a cause of the expansion of the industry in America." Compare Ingvar Svennilson, *Growth and Stagnation in the European Economy* (Geneva: United Nations Economic Commission for Europe, 1954), p. 123.

proposed in the historical literature, emerging as the theory of technological change "embodied" in new capital equipment. And like the interrelatedness argument, it provided what appeared to be a plausible alternative to the hypothesis of entrepreneurial failure.

The Quantitative Evidence for Britain as a Whole. Although mildly fashionable among historians, neither of these alternatives could be considered to have been successful in replacing the hypothesis of failure, because both were introduced in the same nonquantitative way as entrepreneurship itself. The form of argument adopted by both sides in the debate was qualitative isolation of one variable, whether sociological or economic. Although both argued that their variable—entrepreneurship, interrelatedness, slowly growing demand—was sufficient to explain a good part of the apparent lag in technology, only two attempts have been made to show that a favored variable was potent in the British case. One of these, Peter Temin's application of the argument from slowly growing demand to the British steel industry, is reexamined in detail in Chapter 6 below. It will suffice here to note that its impact is too small to explain any substantial British lag in technology. The other, Paul David's application of the interrelatedness argument to mid-Victorian agriculture, is more successful.[23] But his very success in explaining the British lag in adopting the mechanical reaper, on the grounds that the interrelated equipment —namely, the land—was inappropriate to the newer technology, suggests why the argument could have little force outside of agriculture. Farmers as a group must work with the land as they find it, as in the case of mid-nineteenth-century Britain with land plowed into the ridge-and-furrow configuration appropriate to

23. "The Landscape and the Machine: Technical Interrelatedness, Land Tenure and the Mechanization of the Corn Harvest in Victorian Britain," in D. N. McCloskey, ed., *Essays on a Mature Economy: Britain after 1840, Papers and Proceedings of the MSSB Conference on the New Economic History of Britain, 1840–1930* (London: Methuen, 1971).

11

earlier agricultural technology but highly inappropriate to an age of drainpipes and mechanical reapers. Industrialists, however, need not work with the equipment of their predecessors: unlike the farmers, they can as a group move their operations, if they wish, to new factories and mills. The economic explanations of the lag in British technique in the late nineteenth century, in short, share with the sociological explanations a notable lack of quantitative bite.

The prior question whether or not there were in fact any failures to be explained has remained unanswered, and it is on this question that quantitative methods can shed the most light. The index of economic failure implied in writings on the late nineteenth century is the extent of neglect of new techniques, techniques that could have raised the output of textiles, coal, or iron wrested from given amounts of labor, capital, and raw materials and could therefore have yielded higher profits for the entrepreneur who adopted them. Failure can be measured by the size of this output or profit foregone by British entrepreneurs. If it was high—if, that is, the productivity of the British economy was low or was increasing unusually slowly in the late nineteenth century by comparison with earlier experience or with Germany and the United States at the same time—then the discussion can proceed to explanations. If the output foregone was low or nonexistent, however, the hypothesis of entrepreneurial failure would need to be reconsidered.[24]

24. This program of research has already begun, and its preliminary results are for the most part unfavorable to the hypothesis of failure. For the evidence on cotton textiles see Lars Sandberg, "American Rings and English Mules: The Role of Economic Rationality," *Quarterly Journal of Economics,* 83 (1969), 25–43. The slow adoption of the Solvay process in the British chemical industry is examined in Peter Lindert and Keith Trace, "Yardsticks for Victorian Entrepreneurs," and the productivity of the British coal industry in D. N. McCloskey, "International Differences in Productivity? Coal and Steel in America and Britain before World War I," both in D. N. McCloskey, ed., *Essays on a Mature Economy.* An extended review of the issues in the quantitative study of British entrepreneurship is D. N. McCloskey and Lars Sandberg, "From Damnation to Redemption: Judgments on the Late Victorian Entrepreneur," *Explorations in Economic History,* 9 (1971), 89–108.

The evidence on the aggregate rate of growth of productivity in the United Kingdom, the United States, and Germany after 1870 suggests that some such reconsideration might be in order. The test is a stringent one, because there is a presumption that the British economy, being mature, was exploiting available technology better than other countries at the beginning of the period: Britain, in other words, presumably had fewer opportunities to increase productivity by catching up to the leader, for she was already the leader. Nonetheless, down to the very end of the century the rate of growth of productivity in the United Kingdom compared favorably with that in Germany and the United States. Productivity may be measured by subtracting the rate of growth of a composite index of labor and capital inputs from the rate of growth of real national product. The results of this calculation for the United Kingdom from 1860 to 1910 are exhibited in Table 1.

TABLE 1. Growth Rates of Real Gross National Product, Labor, and Capital, and of Total Productivity, U.K., 1860–1910.

| Between the years: | Percentage growth rate per year | | | |
	Real gross national product	Employed labor force	Capital stock	Total productivity
1860–1870	3.3	1.1	1.4	2.1
1870–1880	2.1	1.0	1.6	0.88
1880–1890	2.8	1.0	1.2	1.8
1890–1900	2.2	1.0	1.5	1.0
1900–1910	0.81	1.0	1.5	−0.37

Sources: Details of the estimating procedure are given in Appendix A. An extended discussion of the calculation appears in D. N. McCloskey, "Did Victorian Britain Fail?" *Economic History Review*, 2nd ser., 23 (1970), 455–458. The share of labor was taken to be 0.52 and the share of capital 0.44, as a residual from the share of labor and land. Thus, to illustrate the calculation, between the years 1860 and 1870 total productivity is $3.3 - (0.52)1.1 - (0.44)1.4 = 2.1$ percent per year.

13

During the 1870's, 1880's, and 1890's—the years during which the presumption of entrepreneurial failure was fixed in the minds of Englishmen—total productivity was growing on average at the respectable rate of 1.2 percent per year in the United Kingdom. This is somewhat higher than the rate of productivity change in Germany implied by Walther Hoffmann's estimates of income, capital, and labor from 1880 to 1910 (namely, around 1.1 percent per year) and somewhat lower than the rate in the United States calculated by John Kendrick from 1869 to 1909 (around 1.5 percent per year, or, using an 1889 base to avoid the flaws in earlier data, around 1.3 percent per year).[25] Paradoxically, it is only in the 1900's, when exports and investment had recovered decisively from their disturbing performance of the previous decades, that British performance was poor. Earlier it was good, comparable to that of the period's two most rapidly developing economies. Many observers have agreed with David Landes that "there is no doubt . . . that British industry was not so vigorous and adaptable from the 1870's on as it could have been."[26] Although "industry" is a narrower concept than the national income concept used here and what "could have been" depends on one's beliefs on what was possible, it appears that there is little support for this gloomy view in the aggregate statistics.

The Relevance of the Experience in Iron and Steel. These are fragile foundations, however, on which to erect theories of British success or failure. What is required is detailed confirmation of the result from a sample of British entrepreneurial performance in the late nineteenth century, and the iron and steel industry is the best place to begin. The evidence, both quantitative and qualitative, is

25. W. G. Hoffmann, *Das Wachstum der deutschen Wirtschaft seit der Mitte des. 19. Jahrhundert* (Berlin: Springer, 1965), pp. 827–830 for national income in constant prices, pp. 204–205 for the labor force, pp. 253–254 for the capital stock, and p. 87 for the share of labor in income. J. W. Kendrick, *Productivity Trends in the United States* (Princeton: Princeton University Press, 1961), p. 331.
26. Landes, "Technological Change," p. 559.

ployment available. The resources (including purchases from other industries, excluded from measures of value-added) would continue to be useful, albeit less useful, and the loss to the nation would be only the reduction in their usefulness. One can think of an industry as a collection of techniques that make resources more valuable than in alternative uses. The importance of the iron and steel industry to national income is by this definition the increment arising from these techniques over the next best alternative uses of ore land, coal land, labor, and capital. Although it is clear that the increment has little relation to value-added or employment in the industry, it is not clear what in fact it was in iron and steel. The ease with which it can be measured depends on how evident are the substitutes for the industry's product. This in turn depends on how broad is the historical question at issue. If the question is how important in this sense the Bessemer steel industry was, the substitutes—iron and open hearth steel—are relatively obvious and the magnitude of the increment to national income can be estimated with fair precision. If the question is how important the entire iron and steel industry was, however, the substitutes—stone arches in bridges and wooden parts in machinery—are obscure and the question is correspondingly difficult to answer.

The alternative definition of importance is the one more relevant to the narrower issue of entrepreneurial failure in iron and steel: it is the impact that given magnitudes of changes in productivity or demand in the industry would have on national income. The measure of importance appropriate to this definition is the ratio to national income of the industry's gross output (not value-added or employment, but the value of all sales to other industries). In competitive equilibrium the value of sales is equal to gross cost (the opportunity cost of all inputs, including purchases from other industries), and gross cost is the value of steelmaking resources in their alternative uses, such as house building, coal raising, and bread baking. If demand falls by 1 percent, the resources released for alternative uses are worth 1 percent of gross output, not 1 percent of value-added or employment. Again, if the ineptness of

17

entrepreneurs causes 10 percent more labor, capital, ore, and coal to be required to produce iron and steel, the loss to national income is the value of the additional resources required, namely, 10 percent of the former value of gross output. The national loss in percentage terms, therefore, would be 10 percent of the gross output of iron and steel divided by national income. The ratio of gross output to national income measures the percentage impact of a 10 percent failure of productivity. It is not the shares of value-added or employment in the totals that measure the national importance of a supply failure or a demand reduction in iron and steel but the ratio of gross output to national income.[28]

If all commodities were produced directly for final use the rele-

28. Two qualifications are ignored here, general equilibrium effects and rent. The small size of the industry justifies ignoring general equilibrium effects. Rent on ore land, even in the long run, might appear to be impossible to ignore. An improvement of ironmaking technique, quantity produced held constant, frees ore lands for alternative uses. But they have no alternative use (that is, their reward is rent from the point of view ~f the iron and steel industry as a whole), so there is no gain to n~.ional production on this account. To the extent that it contains rent, then, the value of gross output overstates the real gain to the nation. There are three reasons why this refinement is unimportant. First, domestic ore land of a given quality was not in completely inelastic supply: there would be some real gain from using better ore. Second, only part of the price of ore was rent: the mining labor and capital *did* have alternative uses. Third, the industry faced an elastic supply of foreign ore. The refinement, in any case, would reduce the importance of iron and steel. There is another point about this measure of importance that may be puzzling. It is true that with intermediate production the sum of gross outputs is greater than national product, which implies the apparently paradoxical fact that a 10 percent fall in productivity in all industries results in a more than 10 percent fall in national product. The air of paradox, however, is easily dispelled. A fall in productivity in all industries amounts to throwing away a certain absolute value of resources used directly and indirectly in the industries. When the loss is described by dividing it by gross output, the percent drop in productivity may be 10 percent in each industry; when it is described by dividing it by national product (or the total of value-added), a smaller number than gross output, the result for all industries is larger than 10 percent. The difference is merely definitional. E. Domar provides an algebraic analysis, in the context of input-output analysis, of the gross output weighting of productivity change in "On the Measurement of Productivity Change," *Economic Journal*, 71 (1961), 709–711.

vance of the value of gross output would be transparently clear, for if iron and steel could be eaten, the lost iron and steel could be seen to be a direct loss of national economic welfare in proportion to the industry's sales. The point becomes obscured because iron and steel are intermediate goods. The obscurity attaches to other intermediate goods as well: in common historical usage, coal and transport as well as iron and steel are considered "basic" because they enter into many other products. There must be multiple costs, it is felt, of failures in such industries. The preceding argument shows why this intuition is mistaken. It does not matter at what stage of production a loss from inefficiency occurs: if it costs £5 million worth more resources to produce steel, the loss is the same as £5 million worth lost in the production of domestic service or food retailing. Because the consumers of iron and steel (such as the engineering and construction industries) are not made less efficient by the higher cost of iron and steel, the real loss is not multiplied. In short, there is little warrant for speaking of the iron and steel industry as peculiarly "basic."[29]

The value of gross output, then, is the relevant magnitude for measuring the importance of entrepreneurial failings in iron and steel. In 1907 it was £86 million, eighth among the fifteen census industries and about 5.8 percent of their total gross output.[30] It

29. It is noteworthy that the estimates of social savings attributable to the American railways made by R. W. Fogel in *Railroads and American Economic Growth* (Baltimore: Johns Hopkins Press, 1964) and A. Fishlow in *American Railroads and the Transformation of the Ante-Bellum Economy* (Cambridge, Mass.: Harvard University Press, 1965) involve precisely the same issue. The railroad can be thought of as an innovation that increased the productivity of transportation. Weighting the percentage productivity increase by the ratio of the gross output of transportation to national income (the ratio of importance in the terminology used here) gives the static impact of the railways on national income in percentage terms. There are no multiple effects.

30. The gross output of iron and steel is the value of steel, wrought iron, and cast products sold to other industries plus the estimate in the Census of the factory value of pig iron and other semifinished products exported. The gross outputs of the other industries are also nonduplicative

19

was 4.4 percent of net national income. A difference in total productivity in the making of iron and steel of as much as 10 percent compared with its level under vigorous entrepreneurs, in other words, would have reduced national income less than one-half of 1 percent below what it would have been, or less than one-third of one average year's growth in real national income from 1890 to 1910. In this magnitude of failure in iron and steel took effect over a twenty-year period, the lost income growth due to entrepreneurial laxity would have been about one-fiftieth of 1 percent per year. This is not the stuff of climacterics and massive retardations in economic growth.[31]

Assessing Entrepreneurial Performance in Iron and Steel. If the iron and steel industry cannot by itself be blamed for any great national loss from bad entrepreneurship, it can nonetheless stand as an example of the worst that British entrepreneurs could offer— or so, at least, the industry has been portrayed. Entrepreneurial performance is notoriously difficult to measure, and, in consequence, supporters of the hypothesis of failure in iron and steel have on occasion adopted a curious mixture of agnosticism and certitude in discussing its measurement. P. L. Payne, for example, remarks that "to make any assessment of just how much the relative decline of the British iron and steel industries was due to

---

in the sense that all sales within an industry, estimated in the introduction to each industry's analysis by the Census, are excluded. That is, they too are sales to other industries.

31. A similar argument applies to the four industries taken together that have been indicted for special failures in the late nineteenth century, railways, mining, chemicals, and iron and steel. Gross freight and passenger receipts on United Kingdom railways were about £122 million in 1907 according to the Railway Returns (given in Mitchell, *Abstract of British Historical Statistics,* pp. 226, 229). The gross outputs of the others are given in the Census of 1907, and the total for all four is about £402 million, about one-fifth of the national income. A productivity failure in all of them of as much as 10 percent, then, would reduce income by only 2 percent in 1907, contributing very little to a climacteric in productivity growth.

these entrepreneurial failings is impossible,"[32] then goes on to assess the contribution of entrepreneurial failings high. And Burn, in the midst of a statement of its importance, asserts that "it is not possible to prove or measure . . . [the attachment to routine]; it leaves no infallible testimony and can only be inferred from the narrative."[33] Contrary to this agnosticism, however, entrepreneurial failure *does* leave testimony. The task of this study is to examine the testimony and to reach a verdict.

Of course, there is part of Burn's assertion with which it is difficult to quarrel: no single bit of testimony is infallible when considered by itself. Since German and especially American entrepreneurs were alleged to be superior, the most direct test of the hypothesis of failure, a test performed in the last chapter of this work, would be a comparison of the productivity of the British and the foreign iron and steel industries. If it appeared, as it does, that the British and foreign levels of productivity were equal, British entrepreneurs could be pronounced innocent of the charge of failure. The better rule of evidence is, however, as in the civil law, *unus nullus*: the testimony of a solitary witness on a point is to be treated as no testimony at all. Entrepreneurial performance leaves scattered traces, and the task in what follows is to gather together these bits of evidence, both direct and circumstantial, on its quality. The method is to infer its quality as a residual explanation of the behavior of the industry: when directly measurable influences on the industry's demand or supply cannot explain the industry's performance, what is left to be explained may be attributed to entrepreneurial quality.

32. "Iron and Steel Manufactures," in D. Aldcroft, ed., *The Development of British Industry,* p. 95.
33. *Economic History of Steelmaking,* p. 213.

2   The Market Structure of the

Industry

The intellectual machinery of supply and demand in its simplest form, used intensively later, is applicable only to a perfectly competitive industry. If the British iron and steel industry during the years 1870 to 1913 diverged markedly from perfectly competitive behavior, the range of analytic possibilities in the measurement of entrepreneurial quality would be great and any empirical inquiry correspondingly difficult. A monopolist, to take just one example, does not have a supply curve, making it doubtful that the isolation of the impact of good or bad entrepreneurship on the industry's supply curve has any useful meaning. The traditional escape from this dilemma by bald assumption has attractions, for then the inquiry can continue for those, at least, who are willing to accept the competitive assumption. But it would be more satisfactory to decide at the outset whether or not the assumption was in fact true, the more so as its truth or falsehood sheds some direct light on the hypothesis of failure.

Cycles in Monopoly Power from 1870–1913. There are indications that monopoly power may have been important in the British iron and steel industry. To be sure, the number of firms making pig iron (the chief raw material for finished iron and steel) and cast and wrought iron (the older forms of finished iron) was large,

22

which would presumably make collusion to raise prices above competitive levels difficult. The number of firms making heavy steel products, such as plates, rails, sheets, and girders, however, was small. Furthermore, all branches of the industry were widely dispersed over the country's coal and ore deposits, making the number of firms in each regional market still smaller, small enough for one to suspect that in some regions collusion may have been effective even in pig, cast, and wrought iron. The industry responded to this incentive by engaging in persistent attempts, some of which are known to have been successful, to bring into existence cartels, price agreements, monopolistic mergers, and restrictions on output.

It is clear from this evidence that the iron and steel industry was not perfectly competitive in any unlimited sense. It is equally clear, however, that the industry was not organized into a perpetually effective cartel. British makers of iron and steel broke as well as made collusive agreements, central selling was not instituted in the industry until the 1930's, entry could not be completely blockaded until the industry was nationalized, new firms did enter the industry and moved from branch to branch, and, until the iron and steel tariffs of the 1930's, foreign competition set a limit on monopoly. The question, then, is very much one of *how far* the British industry in the late nineteenth century deviated from competition. Here the traditional evidence concerning cartelistic agreements and amalgamations fails, for it does not tell how effective the agreements and amalgamations were in claiming the rewards of market power. What is needed is some measure of this power.

The natural measure of market power is the proportional excess of price over marginal cost.[1] For a profit-maximizing firm in equilibrium, the higher the elasticity of demand that it believes it

---

1. As proposed by A. P. Lerner, "The Concept of Monopoly and the Measurement of Monopoly Power," *Review of Economic Studies,* 1 (1934), 157–175.

faces the lower will be the proportional excess of price over marginal cost, until, in the limit, a perfectly competitive firm, believing it faces an infinitely elastic demand curve, sets marginal cost equal to price. The four important commodities in the heavy iron and steel industry from 1870 to 1913 were pig iron, wrought iron common bars, heavy steel rails, and steel ship plates. The obstacle to measuring the excess of price over marginal cost for these is that although there is good information on price there is very little on marginal cost. The way around the obstacle is to observe the market price when there is some presumption that pricing *was* competitive and to compare it with the price when there is a presumption that pricing was *not* competitive. In competitive circumstances the price of a product is equal to its marginal cost; subtracting this competitive price from the price observed in monopolistic circumstances will give the excess of the monopolistic price over marginal cost. Although it is some advance over agnosticism, there is, of course, no certainty in this procedure: without direct evidence on marginal cost one can only make probable inferences on the magnitude of monopoly power, moving cautiously from what is known to what is not. It is desirable, therefore, to repeat the procedure in as many ways as possible and to bring to bear related evidence on monopoly in the industry at each step.

One way to apply the reasoning is to compare the prices of products at different dates. The price of a finished product, heavy steel rails, say, rose and fell because of changes in the prices of inputs, changes in productivity, and changes in the degree of monopoly power. The price of the most important input, pig iron, is readily available. The observed ratio of the rail price to the pig iron price will reflect productivity and the degree of monopoly power. As Peter Temin put it in developing this reasoning for the American iron and steel industry in the late nineteenth century, "the troughs of this ratio reflect the prices when the pools broke up and competition was allowed to operate . . . and the peaks of the ratio indicate the level to which the relative price of steel was

24

raised by the action of the pools."[2] The trend in the ratio is an estimate of the trend of productivity in railmaking and can be eliminated to leave the cycles in the degree of monopoly power.[3] The average excess over the minimum ratio of rail prices to pig iron prices will be the average excess of price over marginal cost, an upper bound on the degree of monopoly power in railmaking. Applying this test to steel ship plates, wrought iron common bars, and heavy steel rails shows that for the period 1870 to 1914 taken as a whole the branches of the iron and steel industry producing finished products were reasonably competitive. To summarize the calculation with one number, the deviation of price from marginal cost was no higher than 7 percent on average, and in many cases was lower.

The heavy steel rail trade after 1895 is the one exception to this assertion. The story of the frequent price and output agreements in rails has been told many times, and it is unnecessary in view of the confirming quantitative evidence to recount the rele-

2. *Iron and Steel in Nineteenth-Century America: An Economic Inquiry* (Cambridge, Mass.: M.I.T. Press, 1964), p. 187. An early use of essentially the same method is J. T. Dunlop, "Price Flexibility and the 'Degree of Monopoly,' " *Quarterly Journal of Economics,* 53 (1939), 522–534.

3. These cycles are also cycles in the marginal cost of production, unless the marginal cost curve of steel was perfectly elastic in the short run. Whatever the elasticity of supply of the input, pig iron, the ratio will cycle as output does, so long as there are any capacity constraints on steel production: the price of pig iron (and other inputs available in somewhat elastic supply) will rise when the steel industry increases output, but the marginal cost of steel will rise still further, because of the constraints set by the inelastic factor, capital. If market power rises as output does (a likely event), the effects of rising market power and rising marginal cost will be confounded in the index. Part of the supposed rise in the index of market power will be merely a rise in marginal cost and the index will be an upper-bound measure of market power. On the other hand, if market power falls as output rises (that is, if cartels break up in booms and form in depressions), the index will be, if anything, a lower-bound measure. The evidence on cartel formation in iron and steel suggests that the former characterization is more appropriate: cartels appear to have formed in booms and broken up in depressions, and the measure of monopoly power will be therefore an overstatement.

vant anecdotes here in any great detail.[4] In 1883 the British Rail-makers' Association was formed and soon after combined with German and Belgian makers into the International Rail Syndicate. The stability of published steel rail prices in 1884 and 1885 does suggest that prices were indeed being set by a committee rather than by the market, although it does not tell how far above marginal cost the committee was able to raise prices.[5] This first rail syndicate collapsed in April 1886, with an epitaph in the *Iron and Coal Trades' Review,* that "in this country, combinations among manufacturers to regulate production and prices are seldom successful for long."[6] Until 1893 there is no further evidence of rigid prices in the rail trade. In 1893, however, the reported prices of heavy rails for the three principal producing centers, Cleveland, South Wales, and the Northwest Coast, developed a rigidity that lasted with two interruptions in 1899–1900 and 1904–1905 down to 1914.[7] In February 1896 the British Rail-makers' Association was officially revived and in November 1904 it joined again in an International Railmakers' Association. From January 1906 onwards the quoted price of heavy rails was the same in the three major railmaking districts. In this environment it is not surprising that the *Iron and Coal Trades' Review* com-

4. The basic study is H. W. Macrosty, *The Trust Movement in British Industry* (London: Longmans, 1907). J. C. Carr and W. Taplin retell the story with new material in *A History of the British Steel Industry* (Cambridge, Mass.: Harvard University Press, 1962). Burn and Clapham tell it in less detail.

5. The price data for rails are poor before 1883, but on the Northwest Coast (for which there are rail prices continuously from January 1883 on) during 1884 and 1885 prices are constant for stretches up to 15 months long. See Appendix B for the rail prices.

6. July 9, 1886, p. 55.

7. Rail prices were collected for the three regions for the week nearest the 15th of January, April, July, and October from 1883 through 1913, from the contemporary issues of the (weekly) *Iron and Coal Trades' Review*. From April 1883 through January 1886, therefore, there are 12 successive prices. Six of these 12 are unchanged from the price three months before. From January 1893 through October 1913, 32 out of the 80 prices are unchanged.

plained, in December 1906, that "there has never been, in the whole of the varied and chequered history of the iron trade, such general regulation of prices by combinations of manufacturers, merchants, shippers, and others as at the present time."[8]

In the steel rail trade, then, general regulation of prices was common from its earliest years and persisted. The ratio of rail prices to pig iron prices tells how effective it was. By this criterion, from 1883 to 1895 only one year, 1885, looks convincingly like a year of effective collusive pricing: in that year the ratio of rail to hematite pig iron prices in Cumberland was about 2.33, compared with an average of 1.97 during the four years around it. Moreover, 1885 was a year in which the output of rails was low: the price of rails was high relative to pig iron because the Railmakers' Association was pushing it up, not because rail output was straining against capacity.[9] This is only one year out of thirteen, however, and the ratios during the other years were much closer to the trend. Consequently, from 1883 to 1895, despite the existence of the Railmakers' Association in the first four years, the average excess of price over marginal costs was only about 5 percent, and only about 2.5 percent if the year 1885 is omitted.[10] After the revival of the Association in 1896, however, the situation changed sharply and the ratio of rail to pig iron prices remained on average well above its minimum levels down to the war. The high ratios were not continuous and there is some evidence that monopoly profits attracted new, price-breaking firms

8. Quoted in Carr and Taplin, *History of the British Steel Industry,* p. 255.

9. Rail exports were unusually low during 1884, 1885, and 1886 and the output of Bessemer steel was low during 1884 and 1885.

10. These estimates are derived from the prices in Appendix B. They are for steel rails made in Cumberland from Bessemer pig iron. The estimates for steel rails made in Cleveland are less satisfactory because prices of forge pig iron rather than Bessemer pig iron had to be used. On the other hand, the Cleveland steel prices reach back into the 1870's. Over the periods 1872–1879 and 1883–1895 (1884 omitted for lack of data), the indicated excess of price over marginal costs is about 7 percent. The ratio of rail to pig iron prices is particularly high in 1875, 1877, 1885, and 1886.

into the rail trade from time to time.[11] Despite these episodes, however, the implied average excess of price over marginal cost was high from 1896 to 1914, about 30 percent compared with the 5 percent figure from 1883 to 1895.[12] That is, after the mid-1890's price controls in rails became a good deal more effective than they had been before.

Nothing like this, however, occurred in the other large branch of the heavy steel trade, ship plates, or in the wrought iron bar trade. According to the ratio test, price typically overstated marginal cost by 5 percent as an upper bound between the years 1883 and 1913 in Scottish heavy ship plates. Unlike rails, the years of the highest apparent excess of price over marginal cost in ship plates are also years of high outputs, cyclical peaks of steel shipbuilding: it is probable that the high ratios of plate to iron prices in 1889–90, 1906–07, and 1911, for example, come from strained capacity rather than from increased monopoly power. Although the prices of ship plates in Scotland and Cleveland became, like rail prices, more rigid after the mid-1890's, no marked increase in the excess of price over marginal cost accompanied this rigidity, in contrast to the pattern in rails. Indeed, after 1900 in the wrought iron bar trade, competitiveness appears by this test to have increased. In short, the price ratio test indicates that, except for the rail trade after 1896, which was in any case at this time a shrinking portion of the whole, the iron and steel industry was substantially competitive.

11. Burn, *Economic History of Steelmaking,* p. 276: "In Britain, new firms even forced their way into the heavy-rail trade in the 'nineties and early nineteen-hundreds, though the market was shrinking." He mentions Cargo Fleet, Glengarnock Iron and Steel Company, Bell Brothers at Clarence, and the Shelton Iron and Steel Company. Cargo Fleet and Glengarnock, incidentally, appear to have been the last British makers to join the International Railmakers' Association and the first to leave.

12. The difference is not so great in the imperfect statistics for Cleveland. Here the excess rises from an average of about 7 percent during 1872–1879 and 1883–1895 to an average of 13 percent during 1896–1913.

Monopoly Power among Regions of the Country. The ratio test compares prices at different dates, using the price prevailing in a competitive period as an estimate of the marginal cost. If one is willing to accept the reasonable premise that cartels were for the most part potentially viable only on a regional, rather than national, level a second measure of the degree of competitiveness in iron and steel can be constructed by comparing prices at different locations, using the price prevailing in another part of the market as an estimate of the marginal cost. Whether or not they were themselves formed into a regional cartel, if Middlesbrough makers of ship plates believed that they faced an elastic demand in the Glasgow market, they would have adjusted their shipments so that the price they received from selling there was equal to marginal cost. The price that they received from selling in the Glasgow market was the Glasgow price of plates minus the cost of transportation from Middlesbrough to Glasgow. For a given price in Glasgow, the situation of the Middlesbrough makers of ship plates would have been as shown in Figure 1. If the Middlesbrough makers competed with each other their price would have been P_0, equal to marginal cost. If, on the other hand, they formed an effective regional cartel they would have sold some plates to Glasgow at the Glasgow price minus transport cost (equal to marginal cost) and the rest to Middlesbrough at the higher price, P_1. Of course the Glasgow price minus transport cost might have been so low that the Middlesbrough cartel would have ignored the Glasgow market. That is to say, the Glasgow price minus transport cost is a lower bound on the marginal cost at Middlesbrough. It would have been in fact lower than the marginal cost at Middlesbrough either if the Middlesbrough makers did not sell to Glasgow or if they competed with each other. The excess of the Middlesbrough price (P_m) over marginal cost (MC_m) is, then, at most,

$$\frac{P_m - MC_m}{MC_m} = \frac{P_m - (P_g - T)}{P_g - T}$$

29

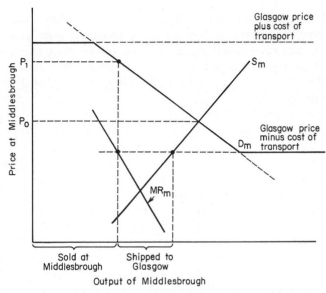

Fig. 1. The Limits on Monopoly Power in a Region

(where T is the transport cost from Middlesbrough to Glasgow).[13]

It is clear from the formula, and in any case obvious, that the ratio of transport costs to the price of the product is crucial in determining the maximum possible excess of price over marginal cost: the makers of products more expensive relative to their transport costs (for example, steel ship plates and common iron bars) would have had less room to charge prices in excess of marginal costs than the makers of cheaper products (for example, steel rails and pig iron), because the cheaper products were protected from outside competition by transport costs. The price of rails at Middlesbrough, for instance, could have been no greater

13. Strictly speaking, regional monopoly is the extreme case only if regional oligopoly would not lead to a greater excess of price over marginal cost. An oligopolistic industry would have lower profits than a monopolized one, but this does not imply that the oligopoly price would be lower than the monopoly price. Less monopoly power does not necessarily lead to less misallocation.

than the price of rails at another town in the rail trade, say Furness, *plus* the transport cost from Furness to Middlesbrough: if it had been greater, well-informed and profit-seeking Middlesbrough buyers and Furness sellers could have benefited by trading. The marginal cost of rails at Middlesbrough, likewise, could have been no less than the price of rails at Furness *minus* the transport cost: if it had been less, well-informed and profit-seeking Middlesbrough sellers could have increased their profits by selling in Furness.[14] The buyers and sellers appear to have been in fact well informed and profit seeking, for prices did stay within these bounds.

The larger was the transport cost in relation to the price of the product, then, the less effective would outside competition have been in restricting regional cartels and the greater would have been the potential excess of price over marginal cost. If rails had been priced in Middlesbrough at their maximum level relative to the Furness price (that is, if P_m, the Middlesbrough price, was set as high as $P_f + T$, the Furness price plus the cost of transport) the formula reduces to

$$\frac{P_m - (P_f - T)}{P_f - T} \leq \frac{P_f + T - (P_f - T)}{P_f - T} = \frac{2T}{P_f - T}$$

For rails this extreme bound is quite high, typically around 30 percent of the price: consistent with the earlier findings on the rail trade, transport costs could have protected a good deal of cartelistic pricing.

Interregional competition was much more effective in the other

14. Notice here again the assumption that Middlesbrough sellers were perfect competitors in outside markets like Furness. If, on the contrary, they had market power in Furness, they would sell in Furness up to the point at which their marginal cost equalled marginal revenue, not price (which would be higher than marginal revenue). In this case, the Furness price minus transport costs would *not* be a lower bound on marginal cost and the formula above would not be an upper bound on the excess of price over marginal cost. This is not a great problem in using the formula, however, because there is usually some market with a roughly equal price over which the particular British maker could not have had market power.

trades, ship plates, for example, as the anecdotal evidence suggests. A manager of the Consett Company on the Northeast Coast spoke in his letters in 1887 as if Scotland was from his point of view an anonymous market in which he was a price-taker. Indeed he was: two years before his company had wrecked a combine of Scottish ship plate makers which had raised the price of ship plates by ten shillings.[15] Again, the ship plate makers tried repeatedly during the 1890's and 1900's to form a national combine, a "united front," as they called it. In September 1904 the Scottish and Northeast Coast ship plate and angle makers agreed not to invade one another's territory. They could not convince Guest Keen's in South Wales to honor the agreement, however, and Guest Keen's wore away the combine price in the West Coast and at Belfast.[16]

Because of regional competition such as this the rail trade alone could be effectively cartelized before 1914 on a regional level. The price of steel ship plates averaged 128 shillings a ton in the northeast of England and 133 shillings in Scotland from 1885 to 1914, while the cost of rail transport between Scotland and the Northeast Coast was about 12 shillings a ton. These prices imply a maximum excess of price over marginal cost in the northeast of about 6 percent.[17] Scotland had a higher price and is therefore more likely to have been, in fact, a region in which there was a local monopoly: the maximum excess for Scotland is higher, about 15 percent. But the low maximum excess of price over marginal cost of the Northeast Coast ship plate trade is duplicated elsewhere in the iron and steel industry. In the Northeast Coast pig iron industry, making a third of British output in 1900, the average excess for 1870–1914 is about 5 percent. Examples such as these could easily be multiplied, for the reasoning implies that any low-price region is likely to have a low maximum excess of price over

15. Macrosty, *Trust Movement*, p. 66.
16. Burn, *Economic History of Steelmaking*, p. 343.
17. By the formula above,
$$\frac{128 - 133 + 12}{133 - 12} = 0.06.$$

marginal cost. Indeed, iron and steel prices in Britain were generally below German export prices at the German border before 1900, so that before 1900 the whole of Britain was in the same relation to Germany as was the Northeast Coast to the rest of Britain, namely, a region unprotected by tariffs in which prices were low and therefore by the logic of interregional arbitrage in the absence of a cartel covering all the market necessarily a substantially competitive region.[18] The regional test, in short, supports the earlier conclusion that competition was broadly effective in the iron and steel industry.

Competition among regions, of course, was not the only obstacle to cartelization in British iron and steel: in the iron bar and pig iron trade competition was enforced by a large number of firms nationally, and often within each region. There were a large number in more ways than one, for the many merchant speculators in the pig iron market needed to be brought into a collusive agreement for it to survive long, and there were as well submarginal puddling and blast furnaces ready to enter and share any high collusive prices. Even without the threat of new entry, further, it would have been relatively easy for one of the seventy-six makers of bar iron in North and South Staffordshire in 1887, for example, to break an agreement without being detected. In ship plates and rails nationally there were many fewer firms—in 1912 only twenty makers of steel ship plates and seventeen makers of heavy steel rails—yet even in these trades the potential entry of new firms was an obstacle to collusion. The fate of price agreements in the ship plate trade reflects this obstacle. The relevant entry conditions are those into a specific *trade* within the iron and steel industry: it was difficult, perhaps, to become a steelmaker, but not difficult to move from, say, ship plates to boiler plates. The Scottish Boiler Plate Association, for example, collapsed in 1903 when a ship plate maker moved into the trade. And even becoming a steelmaker was not impossibly difficult, particularly

18. Burn, *Economic History of Steelmaking*, p. 112, for German and British prices.

in view of the long-lived plant and equipment in steelmaking. Scottish shipbuilders, tiring of high prices for ship plates, reopened in 1913 the Cambuslang Works of the Clydebridge Company, which had been closed in 1907 by the ship plate association to restrict output.[19]

In view of the ever-present danger of new entrants, indeed, it is curious that despite the efforts of the makers the ship plate trade did not become more cartelized after 1895, while the rail trade did. It is possible that it was easier to enter the ship plate trade from related trades than to enter the heavy rail trade, rail-rolling equipment being perhaps more specialized, although the spectacular entrants into railmaking after 1895 do not support this view. An alternative and more convincing explanation turns on the radically different conditions of technological change and of the growth of demand in the two trades after the early 1890's. As will be shown in detail later, the ratios of the prices of rails and ship plates to pig iron reflect technological change in their long-run trends. Although both had stopped falling by the early 1900's, the ratio falls much more slowly in rails than in ship plates during the 1880's and 1890's. Shifting and uncertain costs resulting from unusually rapid technological change before 1900 in ship plates, therefore, would have made agreement even among the few makers difficult. And while ship plate makers had the additional disturbing factor of a continuing growth of demand in the 1890's and 1900's, railmakers did not. British exports of iron and steel for use in railway construction reached their all-time high in the boom of 1887–1890, but the value of new ships and boats exported and the tonnage of ships (especially steel ships) built in the United Kingdom for British citizens or companies continued to grow rapidly during the decades before the war. The stagnant technology and demand in rails permitted agreement among firms to charge a high and stable price, while the rapidly developing technology and demand in ship plates permitted only competition.

19. Carr and Taplin, *History of the British Steel Industry,* p. 259.

Competition and the Hypothesis of Failure. Recognizing that competition and price-taking behavior characterized the milieu of British entrepreneurs weakens the arguments for entrepreneurial failure in a number of ways. For one thing, it makes it difficult to argue from an analogy between the behavior of an individual firm and the behavior of the industry. Firms do indeed grow old and families often do verify the proverb: "Clogs to clogs in three generations." Burn remarks that "the corrupting influence of fortune . . . was of course bound to be greater in the second and third generations of an industrial dynasty than in the first" and applies this notion to the British industry in the 1890's.[20] "In many firms," writes Landes, "the grandfather who started the business and built it up by unremitting application and by thrift bordering on miserliness had long died; the father who took over a solid enterprise and, starting with larger ambitions, raised it to undreamed-of heights, had passed on the reins; now it was the turn of the third generation, the children of affluence . . . [who] worked at play and played at work."[21] And Charles Kindleberger concludes: "The passage of generations may often be the worst enemy of the family firm."[22] It does not follow, however, that it is the worst enemy of a competitive industry. Each firm may trace a pattern of vigor over its history, but at any one time a competitive market directs demand to the more vigorous firms, attenuating the corrupting influence of fortune, the playful propensities of the children of affluence, and the other hazards of the passage of generations. Further, the increasing prevalence of sons—presumably indolent sons—in the industry is easily exaggerated. The iron and steel industry, after all, was not born in 1870: dynasties existed in the industry before 1870 and there is evidence that there was no dramatic increase in the proportion of men recruited from the dynasties after 1870.[23] It is plausible, then, that compe-

20. *Economic History of Steelmaking*, p. 300.
21. Landes, "Technological Change," p. 563.
22. *Economic Growth in France and Britain, 1851–1950*, p. 132.
23. Charlotte Erickson, in her *British Industrialists: Steel and Hosiery, 1850–1950* (Cambridge: Cambridge University Press, 1959), found that

tition did indeed set limits on the impact of the shortcomings of individual families and firms.

The static effect of directing demand to the more vigorous families and firms is not the only relevant consequence of a national, price-taking market for iron and steel, for there is some reason to believe that a competitive industry innovates faster than a monopolistic one. There is, to be sure, also some reason to believe the opposite assertion: the debate among economists on this issue is far from closed. On one point, however, the findings on competition in the industry do speak unambiguously on the mechanism of innovation in iron and steel. Some have argued that the impetus to innovation in Britain was inhibited by slowly growing output. Asserting that there were cost advantages to larger-scale steel plants, for example, Burn asks, "Why had no British firm in the early nineteen-hundreds set out successfully to obtain these advantages to the full? In answering this question a vital distinction between British and rival conditions must be borne in mind. Great new plants could emerge in Germany and the States without other plants being stationary, let alone shrinking or disappearing; in Britain that was possible to a very limited degree, if possible at all . . . German and American makers had far larger and more expansive home markets than the British."[24]

Connecting the size of a market to the size of firms within the

---

the percentage of steel manufacturers whose fathers were partners, owners, or directors in the industry when the manufacturers *began* their careers was 28 percent for men holding office in 1865, 31 percent for 1875–1896, and 36 percent for 1905–1925 (p. 12). From her description of how the sample was gathered, it is apparent that the probable sampling error in each of these figures is high. The increase between 1865 and 1905–1925, therefore, is probably insignificantly different from zero. The percentage whose fathers were iron and steel manufacturers at the time of the sons' *birth* fell in the period from 24 percent in 1865 to 19 percent in 1905–1925 (p. 231). Again, the fall in this figure, like the rise in the other, is probably insignificantly different from zero.

24. *Economic History of Steelmaking,* p. 240. It is difficult to reconcile his argument here with his statement on the page preceding it that the scale of British open hearth plants (these are the plants he is speaking of) was *larger* than the scale of both German and American plants.

market is a popular undertaking, but fraught with difficulties. Burn realizes this and seeks to avoid the difficulties by postulating segmented local markets for iron and steel. If the markets had been sufficiently segmented, a single firm would indeed have avoided great new plants in the face of stagnant demand, realizing, as Burn suggests, that with larger output its price in the local market would fall. The market, however, was not segmented but was a national market of price-takers. Leaving aside the question why it would matter to an individual firm that others were "shrinking or disappearing," a price-taking firm would have little incentive to view the limitation of the total demand in the market for the product it produces as a relevant limitation. On the premise that the firm understood its situation, one would suppose that it would decide whether or not to expand on the assumption that it could sell all it made. On the other hand, a cartel, or one firm in a severely segmented market, might well take into account its effect on the price and therefore might well in similar circumstances innovate less when the innovation required larger output. Firms in iron and steel, however, as was shown above, were forced to act on the whole as price-takers, making it less plausible (although possible, if entrepreneurs misunderstood their situation) that the level and rate of growth of demand in Britain worked in the manner Burn supposes on the impetus to innovation.

Finally, in a more general vein, a competitive market rewards diligence and penalizes sloth by simultaneously raising the profits of efficient firms and reducing those of inefficient firms: slothful firms not only produce less output than they would if they reformed themselves, but they achieve lower profits for their owners and higher profits for the owners of their efficient competitors, who receive a rent for the differential diligence of their employees. That is, competition raises the relative price of the pleasures of sloth, inflicting the cost on the owners of the firm, and making sloth a less viable form of behavior. The owners have an incentive to change managers, for with better men they could earn the high returns of diligence. There is no need to postulate calculating rationality on the part of all owners of iron and steel firms for this argument to have force. Indeed, the argument applies to a monopolized market as well, although the relative incentive is less. All that is necessary is that there exist potential owners who see the higher profits of better management and that ownership can be sold. The perspicacious men would attempt to buy out the old owners at mutually attractive terms and would sometimes, at least, succeed. The evidence suggests that this mechanism of entry operated in the iron and steel industry: there were frequent sales of iron works, advertised in the *Iron and Coal Trades' Review* and other trade newspapers; stocks in many iron and steel firms were listed on the London exchange, and there were local stock markets as well; some mergers in the industry can be interpreted as take-overs of inefficient firms by efficient firms with plans for the capital equipment and labor force of the merged firms; and managers, the key commodity in this market, were mobile between firms. Sloth, in other words, could be bought as well as driven out.

These, then, are the difficulties the market structure of the iron and steel industry creates for the hypothesis of entrepreneurial failure. The difficulties amount to the observation that in a competitive, open, and unencumbered market resources tend to find their way into the hands of men with the best ideas of how to use them. The argument does not, of course, refute the hypothesis, for it gives reason to believe only that resources in iron and steel found their way into the best hands available, not that the available hands were skilled. It does say that whatever arguments are made for or against the hypothesis must give due weight to the competitive environment of the industry.

3 The Industry's Consumers and the Industry's Growth

The industry's competitive market structure reduces the problem of isolating the influence of entrepreneurship to one of isolating other influences on competitive supply and demand. The growth of output in the late nineteenth century appeared to warrant the most gloomy descriptions of Britain's loss of industrial supremacy, and it was a small step from this observation to the assertion that entrepreneurs in iron and steel had been insufficiently vigorous in moving out the industry's supply and demand curves. The output of the industry is most easily measured by the quantity of pig iron produced, pig iron being the raw material for all its products.[1] British output of pig iron was overtaken by

1. A conceptually better index of output (which is, however, difficult to implement for international comparisons) would be the money sales of iron and steel products to other industries in constant prices. Using pig iron output probably slightly overstates German and American output of products in constant prices relative to British output and understates the rate of growth of all three relative to the ideal: Britain produced more of the highly fabricated (and therefore more expensive) products, such as tin plate, and all three countries tended to produce more such products as time passed, rails and iron castings giving way to plates, sheets, and small sections. Imports of pig iron to be made into finished products are not included, but their inclusion would not materially alter the results. British imports of pig iron were on the order of 1 percent of output. Imports, primarily from Britain, were more important to the United States and

the American industry around 1890 and by the German industry around 1905. From the early 1880's to the first World War American output grew at about 7.0 percent per year and German at 5.7 percent per year, while British output grew at the miserably slow pace of 0.93 percent per year, with an even worse record in exports.[2]

The Slow Growth of Demand. Historians have been willing to attribute at least some of the difference in growth rates to forces affecting the industry's demand and supply other than entrepreneurial performance, as when, for example, comparing the United States with Britain, J. H. Clapham reminds us that "half a continent is likely in course of time to . . . make more steel than a small island."[3] The question of how much weight should be given the slow growth of output as prima facie evidence of entrepreneurial failure, however, is a quantitative question and warrants a quantitative answer. When entrepreneurs fail, supply curves do not move out as fast as they could. Limits, therefore, can be placed on the significance of entrepreneurial failure on the side of supply by asking a related question: How much of the slow growth of British pig iron was attributable to a slower rate of growth of the British supply curve? Or, to put it the other way, how much of it was attributable to demand?

The answer depends on which characterization of demand is

---

especially to Germany. Including imports would probably tend to lower slightly the German and American growth rates relative to the British, because the importance of imports fell with time.

2. These figures are continuously compounded growth rates from 1876–1885 to 1904–1913 (ten-year averages are used to accommodate the business cycle). The source for Britain is Mitchell, *Abstract of British Historical Statistics,* p. 132; for the United States, American Iron and Steel Institute, *Annual Statistical Report, 1918* (New York: A.I.S.I., 1919), p. 9; and for Germany, National Federation of [British] Iron and Steel Manufacturers (in later years, British Iron and Steel Federation), *Statistics, 1930* (London: N.F.I.S.M., 1930), p. 134.

3. *Economic History of Modern Britain,* vol. III, p. 122.

appropriate.[4] If there are no transport costs, tariffs, or other impediments to the flow of iron and steel products in the world, all suppliers, including Britain, face the entire world's demand and identical rates of growth of that demand. That is, if it costs a manufacturer of iron in Sheffield no more to sell in Pittsburgh or Essen than at home, output is divided up entirely according to the ability to supply the whole market (Sheffield, Pittsburgh, and Essen together), and it does not matter for the rate of growth of his demand where a manufacturer is located. Under such conditions, therefore, a lower rate of growth of output in Sheffield would be evidence that the supply curve of Sheffield had risen relative to the others' (assuming equal elasticities of supply), and critics of the British industry would be justified in taking the growth of output as an index of the growth of supply. With no impediments, in other words, all of the difference between British and foreign growth rates would be attributable to rises in relative costs, some of which might be attributable in turn to failures of management.

On the other hand, if there are prohibitive impediments to the free flow of products, Sheffield, Pittsburgh, and Essen are each separate markets and a lower rate of growth may indicate merely slower growing demand in the vicinity of Sheffield. It is clear that there were significant, if not prohibitive, impediments, the most obvious and important being between home and foreign markets. The British industry was made to a great extent dependent on the growth of its home market by the costs of transporting iron and steel to foreign markets and, if transport costs provided insufficient protection to the foreign iron and steel industries, by tariffs. And the demand curve on which it depended moved out relatively slowly, as is apparent even without information on the

4. The model here is implicit in Peter Temin's "The Relative Decline of the British Steel Industry, 1880–1913" in Henry Rosovsky, ed., *Industrialization in Two Systems* (New York: Wiley, 1966), esp. pp. 143–149. The conclusions here on the role of demand are similar to Temin's; but see Chap. 6 below.

41

elasticities of the supply and demand curves involved: consumption of iron grew only 1.4 percent per year in Britain after the 1870's, while growing three or four times faster in Germany and the United States. It is reasonably clear, furthermore (although the issue will be touched on again later), that this contrast did not indicate a relative failure in marketing, a failure of British entrepreneurs in comparison with their foreign rivals to encourage home consumption. Iron and steel was used primarily for investment goods, and real net investment in Britain grew in the four decades after the boom of the early 1870's at only 1.25 percent per year, while growing in America and Germany at over 4.0 percent per year.[5] If factories, houses, and railways are not being built, iron is not demanded.

The significance of Britain's slowly growing market for iron and steel at home can be illuminated by a crude statistical experiment. From 1876–85 to 1904–13 the British share in both the home market and the world export market fell, indicating that the British supply curve facing these markets did not move out as fast as did foreigners' supply curves: the British share of British consumption fell from 96 to 78 percent and the British share of world exports from 73 to 34 percent.[6] Suppose, however, that

5. Gross investment would show a still sharper contrast. The British (U.K.) comparison is between 1876–1885 and 1904–1913, from C. H. Feinstein's series of net domestic fixed capital formation in 1900 prices (given in Mitchell, *Abstract of British Historical Statistics,* pp. 373–374). The German comparison is for the same years, from Walther Hoffmann's series in his *Das Wachstum der Deutschen Wirtschaft,* p. 257. The American is for 1877–1886 and 1902–1911, from Simon Kuznets's series (minus net inventory accumulation and increments to claims on foreigners to preserve comparability with the others) given as Series F 139 in United States Bureau of the Census, *Historical Statistics of the United States* (Washington, D.C., 1960), p. 144.

6. The shares are based on pig iron equivalents of output, constructed on the assumption that one ton of finished iron and steel requires one ton of pig iron. It is sometimes assumed in the literature that finished products require more pig iron than their weight, but this is probably incorrect: there is no chemical basis for the assumption, scrap was recycled, and there was an approximate equality between world production of pig iron and world production of finished products of iron and steel. British sales

Britain's supply curve had in fact moved out as fast as did foreigners'; in other words, suppose that Britain had retained her initial shares in supplying British consumption and world exports of iron and steel. Had this been the case, surely, less would have been heard of British failure. It was not so much the more rapid growth of the German and American industries that disturbed Englishmen (after all, as industrializing nations both were devoting a progressively larger share of their rapidly growing resources to the production of iron and steel) as the declining share of Britain in markets that she could have been expected to face on equal terms with foreigners. No tariffs protective of foreign industries prevented the British iron and steel industry from supplying steel to shipbuilders in Yorkshire or to railroads in Latin America.

Had Britain succeeded to this extent, then, German and American exports and British imports would have been lower and British exports and supplies to customers at home higher than they were in fact. British output would have grown faster and foreign output slower, the extent of the relative acceleration of British growth being a rough measure of the supply effects (among them entrepreneurial failure) by themselves. The calculation is straightforward. Ignoring the impact of a more rapidly expanding British supply on world prices and world demand (which would not change the orders of magnitude in the comparison), the reshuffling of world demand under the hypothetical conditions would have raised the British rate of growth of pig iron output from

_____

to Britain, therefore, are estimated as the output of pig iron minus the tonnage of British exports of all iron and steel. The statistics of British imports and exports follow the Trade and Navigation Accounts, as given in Mitchell, *Abstract of British Historical Statistics,* pp. 132, 141, 147; world exports are the estimates of Burn, *Economic History of Steelmaking,* pp. 78, 330, for Britain, Germany, and Belgium for 1876–1885 and these countries plus France and the United States for 1904–1913. Total British consumption on this basis was 4.34 million long tons in pig iron equivalents annually on average in 1876–1885 and 6.43 million tons in 1904–1913; world exports were 4.45 million tons in 1876–1885 and 12.1 million tons in 1904–1913.

0.93 to 2.5 percent per year, and would have lowered the American rate from 7.0 to 6.9 and the German from 5.7 to 4.7 percent per year. Notwithstanding a substantial improvement of British performance on the side of supply, in other words, there would still have been a large gap between the British and foreign rates of growth. Nearly half of the actual Anglo-German gap and nearly three-quarters of the Anglo-American are attributable by this accounting to factors other than Britain's falling share in home and export markets. The slow rate of growth of the markets to which Britain was committed accounts for much of the contrast in rates of growth. Britain's relative decline, it would appear, was to a large extent inevitable: with a slowly growing market at home and with limited access to rapidly growing markets abroad, the industry in Britain could not have grown nearly as fast as it did in America and Germany.

The Substitution of Steel for Iron. The slow growth of demand goes a long way towards explaining not only the relative stagnation of the industry's output as a whole but also the relatively sharper deceleration at the end of the nineteenth century in the growth of British Bessemer and open hearth steel, as distinct from cast and wrought iron. This deceleration seemed especially ominous to contemporaries, for it appeared to indicate that in the most modern branches of the industry Britain was losing ground at an increasing rate. Early in the steel age, of course, the rate of growth was high in all countries, slackening as steel came to dominate iron in the making of rails, bridges, buildings, and ships. The rate of growth of steel output in America fell from the level of 37 percent per year averaged during 1868–1876 to 14 percent per year during 1876–1889. Britain's rate of growth was never far behind in these years: for comparable periods the rates were 33 and 11 percent per year, a difference of only 3 or 4 percentage points.[7] After 1890, amid expressions of alarm over Britain's sud-

7. The British figures are for 1863–1873 and 1873–1889: the earlier initial year and the break at 1873 rather than at 1876 better reflect the

den deceleration, the difference widened to 6 percentage points. While steel output in America grew at 9.3 percent per year from 1889 to 1913 (and in Germany at an almost identical rate), in Britain it grew at only 3.2 percent per year.

The widening difference, however, indicated no special failure on the part of the British industry, but merely the narrowing opportunities for substituting steel for iron. The share of steel in the output of finished iron and steel in Britain nearly quintupled in the sixteen years between 1873 and 1889, from 11 to 50 percent; in the succeeding twenty-four years between 1889 and 1913, of course, it could not match this performance, as a matter of arithmetic: the share rose to 84 percent by 1913, only two-thirds above its level in 1889. The growth of steel was made progressively more dependent on the growth of output as a whole, which was rapid in America and Germany and slow in Britain. The very success of the steelmakers in bringing their product to common use, in short, exposed them after 1890 to the limitation imposed by Britain's slowly growing demand.[8]

---

appropriate comparison between the American and British experience. Output of steel for the United Kingdom is given in the National Federation of Iron and Steel Manufacturers, *Statistics, 1930*, p. 9; for the United States in American Iron and Steel Institute, *Statistical Report, 1918*, p. 23 and Temin, *Iron and Steel in America*, p. 270.

8. The argument can be given a more precise form. The total output of semifinished iron and steel (Q) is the output of steel ingots and castings (Q_s) plus the output of semifinished iron (Q_i, consisting of wrought and cast iron). It follows that the rate of growth of steel output is identically equal to the sum of the rate of growth of total output and a term measuring the rate of substitution of steel for iron (variables with asterisks are rates of change): $Q_s{}^* \equiv Q^* + (Q_i/Q_s)\ (Q^* - Q_i{}^*)$. The rate of growth of semifinished iron and steel (Q^*) can be measured by the output of pig iron minus the exports of pig iron (a more subtle measure would include scrap): this grew at 1.3 percent per year from 1870 to 1913. Output of semifinished iron, estimated as a residual by subtracting steel output from the total, declined at 2.9 percent per year. Had these two rates of growth remained constant before and after 1889, then, the implied rate of growth of steel output would have been $Q_s = 1.3\% + (Q_i/Q_s)\ (1.3\% + 2.9\%)$. From 1873 to 1889 the value of Q_i/Q_s was on the order of 3 or 4, as an average, yielding a rate of growth of steel output of 14 or 18 percent; from 1889 to 1913, in contrast, its value was on the order

The arithmetic of the substitution of steel for iron, then, dominated the development of the British steel industry. Nonetheless, to the extent that it could manipulate the speed with which iron gave way to steel before 1890 (after which the arithmetic became increasingly constraining), it could influence the rate of growth of its output. The substitution of steel for iron went forward apace in the 1870's and 1880's, yet there remains the suspicion that it could have gone forward faster: it is sometimes argued that through a failure of marketing in overcoming the conservativism of their customers, the steelmakers failed before 1890 to nurture sufficiently the demand for their product. In this period before the slow growth of total demand became limiting, steelmakers had an opportunity to increase demand by pressing the virtues of their product with consumers. The episode provides a test of the vigor of British entrepreneurs in influencing demand.

The test is most stringent for the provision of materials for shipbuilding, a branch of the iron and steel industry to which Britain, by international standards, was unusually committed and for which the case for a failure in marketing has seemed to observers of the industry to be most plausible. In a few years in the mid-1870's steel became the dominant material in British railmaking, yet not until the end of the 1880's was a similar position reached in the making of ship plates and angles. Hull plates are the primary use for metal in ships: these must be strong to resist strain, large in size to reduce the number of seams and rivetholes, and malleable to permit easy shaping to the curves of ships. In the qualities of strength, size, and malleability modern steel is definitely superior to wrought iron. As it was put to the British Institution of Naval Architects in 1880. "A material which is stronger than the best iron, and which will double up cold without fracture, has very great charms for the practical shipbuilder."[9] Yet

---

of 0.6, yielding a rate of growth of 4 percent. The mere fall in the ratio of iron to steel output, in other words, is sufficient to explain the sharp deceleration in the growth of British steel after 1889.

9. *Transactions of the Institution of Naval Architects,* 1880, p. 209.

not until 1876 was any significant tonnage of ships built of steel and not until 1885 did the steel exceed the iron tonnage. As early as 1875 an eminent naval architect had remarked that "engineers and shipbuilders are, I think, generally of the opinion that steel must eventually displace iron in shipbuilding, both for hull and machinery. I do not fully understand why the rate of progress towards this desirable end is so slow."[10] The question is whether, as Clapham and others have alleged, this represented "sheer conservatism in the shipyards"[11] and in the steel plants.

A discussant of the 1875 paper made the two essential points in response: "The price of steel must be very materially reduced, even if it is made more reliable than it is at present, before it can be largely adopted in mercantile shipping."[12] Within a few years after 1875 a reliable steel, made by the Siemens open hearth process, did come on the market, and, as improvements in its manner of making during the late 1870's and 1880's reduced its price, shipbuilders did rapidly adopt it. The story is chemical and economic, with little room for "sheer conservatism" on either side of the market.

The chief chemical obstacle to using steel in ships was brittleness. A great many elements make steel brittle, either when the steel is hot and being run through the rolling mill (sulphur and oxygen) or when it is cold (carbon, phosphorous, and nitrogen). Before the steel age, wrought iron, which has little embrittling carbon, had come to dominate cast iron, which has much carbon, because it held up better under the strains of movement and shock in railways and shipping. Cast iron was an impossibly brittle material for ships, and before 1875, because embrittling elements were introduced by its methods of manufacture, Bessemer steel was in a similar position relative to wrought iron. The value of Bessemer's original process was that it cheaply removed carbon from pig iron, leaving a malleable metal like wrought iron. The

10. *Transactions of the Institution of Naval Architects,* 1875, p. 131.
11. Clapham, *Economic History of Modern Britain,* vol. II, p. 62.
12. *Transactions of the Institution of Naval Architects,* 1875, p. 145.

process could not, however, remove sulphur and phosphorous, traces of which embrittled the metal. Consequently, steelmakers learned at an early date to use only pig iron free from sulphur or phosphorous. The embrittling effects of excess oxygen were also recognized and circumvented early. Soon after Bessemer announced his process, Robert Mushet and others discovered how to remove the excess oxygen produced in Bessemer's original metal by melting it together with a manganese-iron compound called spiegeleisen. The difficulty with this procedure was that to entirely avoid brittleness, steel in ships had to be exceptionally "mild," or low in carbon, and spiegeleisen contained much carbon; in avoiding oxygen, steelmakers reintroduced carbon and were unable, therefore, to reach the high standards of malleability required of shipbuilding metal.[13] By 1875, however, with the commercial development in France of a high-manganese and low-carbon compound called ferromanganese, it became possible to produce steel plates sufficiently mild to be used for shipbuilding.[14] Ferromanganese was applied to both the Bessemer process and the recently developed open hearth process for making steel, and for a time both Bessemer and open hearth plates were used for shipbuilding.

Mysteriously, the Bessemer variety of the new steel, although it had as low a carbon content, continued to fail breakage tests, while the open hearth variety did not. Siemens, the inventor of

13. Steel had been used in ships during the American Civil War for running the blockade of Confederate ports. As steel was stronger than iron, less weight was needed for the same carrying capacity, and the resulting additional speed was worth the high price (500 shillings a ton for steel ship plates in 1864 compared with 165 shillings for iron) and unreliable performance of steel. Aside from such profitable and romantic uses, the early steels proved unsatisfactory for ships.

14. The French role was, it seems, to improve the production of ferromanganese until it was cheap enough to be used in preference to spiegeleisen. William Henderson of Glasgow invented ferromanganese, but the company formed to make and use it in 1863 failed. See Lloyd's Register, "Extended Report on Steel for Shipbuilding," *Transactions of the Institution of Naval Architects,* 1877, pp. 385–393, and Burn, *Economic History of Steelmaking,* pp. 52–54.

the open hearth, and many others after him attributed the superiority of his process to the quality control possible with its slower pace; in 1876 he argued that "we are not limited to a blow of ten minutes or a quarter of an hour, but the steelmelter, like a *chef de cuisine* in a French restaurant, takes his sample out, not to taste, but to test."[15] The truth of the matter seems to be, as has been argued by Peter Temin in his history of the American industry, that Bessemer steel contained by the nature of its manufacture still another embrittling impurity, nitrogen. Temin is mistaken when he says that "this factor was undiscovered by the turn of the century,"[16] for in 1889 the British *Journal of the Iron and Steel Institute* summarized in a prominent position an article on nitrogen in steel published the year before in a Swedish metallurgical journal:

'Why is the ingot iron produced in the open hearth of a better quality than that produced by the Bessemer process?' It may, perhaps, be suggested that the Bessemer metal is far less homogeneous in its character than is the open hearth metal, but the great difference that may exist in the percentage of nitrogen dissolved in the respective metals is generally overlooked. In the Bessemer process nitrogen [in the atmosphere] is blown through the molten iron, whilst in the open hearth process the metal is protected from the atmospheric action by a layer of slag . . . [T]he hardness . . . [increases] with the percentage of nitrogen . . . [T]he bad effects of an overblow in the Bessemer converter . . . [are due] to the absorption of nitrogen.[17]

Temin is correct, however, when he argues that steel users based their judgments of steel's qualities not on chemical analysis but on experience and on the reputation of the metal arising

15. *Transactions of the Institution of Naval Architects,* 1876, pp. 149–150.
16. *Iron and Steel in America,* p. 151.
17. *Journal of the Iron and Steel Institute,* 1889, pt. I, pp. 283–284. This quotation is from a summary of H. Tholander's article in *Jernkontorets Annaler,* vol. 43, no. 7. The *Journal* summary is the earliest citation concerning nitrogen in steel in the *Journal's* index. Tholander, therefore, was probably the first to make the point. The first full-scale treatment in the British literature is a long article by Harbourd and Twyman in the *Journal,* 1896, pt. II.

from experience. Nitrogen was the last of the embrittling elements in steel to be discovered and the last to enter the ken of steelmakers. Indeed, in 1909 a partisan of the nitrogen factor was vexed that "the influence of [nitrogen] . . . has, unfortunately, not yet been sufficiently appreciated."[18] Nonetheless, shipbuilders and other consumers of mild steel preferred open hearth to Bessemer steel, and the chemical fact demonstrates that their preference was no irrational whim. The strength of the preference for open hearth over Bessemer steel is apparent in the available information on the output of ship plates made from the two. In 1881 and 1885, for example, in years before and during the major shift from iron to steel shipbuilding, the quantity of Bessemer ship plates and angles was approximately one-fifth of the open hearth quantity.[19] In shipbuilding, then, the substitution of steel for iron was a substitution of open hearth rather than Bessemer steel and awaited the invention of ferromanganese and the open hearth itself.

The substitution did not occur, however, until the late 1880's. The availability of malleable open hearth steel was not in itself enough to drive iron from the market, for the second obstacle to steel's use, its high price, had first to be removed, as was clear to I. L. Bell in 1886: "[I]t may be asked what has retarded the use of steel for shipbuilding . . . The delay must be ascribed to the difference in price."[20] The experience of the industry in Scotland is typical: the price of open hearth ship plates relative to wrought iron ship plates did fall sharply in the 1880's and the

18. W. Giesen, "The Special Steels in Theory and Practice," Iron and Steel Institute, *Carnegie Scholarship Memoires,* 1909, pt. I, p. 1.

19. The figures for the distribution of output are from the annual statistical reports of the British Iron Trade Association in each year. Bessemer plates and angles were 22,000 tons in 1881 and 56,000 tons in 1885, the open hearth plates 100,000 tons and 253,000 tons. The reporting bias in these figures is probably not high. For example, the total of the open hearth products reported is in proportion to the open hearth ingots produced in the two years.

20. *The Iron Trade of the United Kingdom* (London: British Iron Trade Association, 1886), pp. 52–53.

fall did correlate with the rise in the share of steel tonnage in total metal ship tonnage, as Table 2 demonstrates.

TABLE 2. The Steel/Iron Price Ratio and the Share of Steel Tonnage in Scottish Shipbuilding, 1880–1890.

Year	Price of steel ship plates ÷ price of iron ship plates	Share of steel in net tons of ships built in Scotland
1880	1.43	0.15
1881	1.48	.14
1882	1.39	.25
1883	1.36	.31
1884	1.34	.44
1885	1.21	.39
1886	1.24	.62
1887	1.18	.75
1888	1.06	.91
1889	1.06	.96
1890	1.06	.96

Source: The steel prices are given in Appendix B. The iron price is a rough estimate, namely, the price on the Northeast Coast of England of iron ship plates plus twenty shillings (which is roughly the difference between Scottish and Northeast prices in the 1870's). The shipbuilding tonnages are, of course, shipbuilding units (essentially volume measures), not measures of weight; they are reported in the *Annual Statement of the Navigation and Shipping of the United Kingdom,* Sessional Papers, annually. The total includes sailing ships and excludes wood and composite ships. Scottish data alone are used here only to sharpen the test. Net tons are used because gross tons are not given for sailing ships.

The significance of the correlation depends on the elasticity of substitution between steel and iron: if shipbuilders treated iron and steel as good substitutes, the fall in the relative price of steel would be enough by itself to explain the move to steel; if they treated them as poor substitutes, however, the fall in price would have little impact on their relative consumption and one would need to invoke changes in taste, whether autonomous or induced by steelmakers, to explain the move in its entirety. The evidence is not beyond reasonable doubt, but it does suggest a high elasticity and an absence of changes in taste. The ideal test, which cannot be implemented fully, hinges on statistical estimation of the elasticity. If the elasticity of substitution, σ, whether high or low, was constant and if shipbuilders were in competitive equilibrium, then the elasticity could be estimated by fitting a linear equation in which the proportional change in the ratio of iron used to steel used, $(I/S)^*$, is dependent on the ratio of the prices of iron and steel, $(P_S/P_I)^*$, because the elasticity under these conditions will be the slope of the equation. In other words, if one fits the equation $(I/S)^* = a + \beta (P_S/P_I)^* + \epsilon$ the fitted parameter β will be σ.[21] Using this ideal as a guide, the data suggest a high elasticity, on the order of ten, which is high enough for the mere change in prices to explain the shift to steel, without recourse to a putative change in tastes. The contemporary engineering literature confirms this impression: numerous calculations were made of the comparative tonnages of metal required to build a steel or iron ship, allowing for the lightness of steel plates for a given strength, and the calculations assumed that at the appropriate ratio of equal costs of the two, generally around 1.2 tons of iron to 1.0 ton of steel, shipbuilders should be indifferent between

21. This assertion is derived from the mathematics of the constant-elasticity-of-substitution production function, in particular from K. J. Arrow, H. B. Chenery, B. S. Minhas, and R. M. Solow, "Capital-Labor Substitution and Economic Efficiency," *Review of Economics and Statistics,* 43 (1961), 225–250, equation 20. The intercept of the equation, a, can be shown to be equal to $\sigma(b/a)^*$, where a and b are parameters (in the constant-elasticity-of-substitution function) that reflect the level of the marginal products of steel and iron; that is to say, in this context the intercept of the fitted equation reflects the change in the relative usefulness of steel and iron in the minds of shipbuilders. A more refined experiment (the data are too crude and too sparse to make it worthwhile pursuing for the problem at hand) would test whether this intercept was significantly different from zero. If σ is high, a wide range of values of $\sigma(b/a)^*$ around zero are consistent with $(b/a)^*$ itself approximately equal to zero.

the two materials.[22] The ratio would have varied from place to place, depending on the local costs of steel and iron, and from one ship design to another, but it is significant that the shift to steel in Scotland did occur when the ratio came into the range of 1.2 tons of iron to 1.0 ton of steel. In short, it was the fall in the price of a material whose virtues were already widely recognized, not an induced or autonomous shift of sentiment in favor of it, that accounts for the ultimate adoption of open hearth steel in shipbuilding.

The special requirements of the shipbuilding industry, then, made the open hearth the characteristic piece of equipment in the British steel industry after 1890. Britain continued, of course, to make Bessemer steel as well, but this material was closely associated with the making of rails, primarily an export product with a stagnant demand. In depressed years a third, more usually well above half, of Bessemer products by weight were rails, largely exported.[23] Bessemer steel production in Britain reached its all-time peak in 1889, as did rail exports, with 60 percent of the nation's production of steel ingots and castings coming from Bessemer converters.[24] Thereafter, while continuing to rise in America, Bessemer output in Britain fell irregularly, being overtaken by the open hearth in 1894 and falling to 21 percent of ingot and casting production by 1913.

The contrast with the American experience is instructive. Output of Bessemer steel ingots and castings in America reached its

22. For example, in the annual report of the British Iron Trade Association for 1882 (pp. 58–59) "data furnished by a leading firm of shipbuilders" indicated that a typical iron ship requiring 723 tons of iron could be duplicated with a steel ship requiring 524 tons of steel and 106 tons of iron, implying that the two ships would have had the same materials costs when 1 ton of steel cost the same as 1.18 tons of iron.

23. These figures are from the annual reports of the British Iron Trade Association, for the years 1877–1880 in the report for 1880 (p. 29), 1892–93 in the report for 1893 (p. 19; there was serious underreporting of rail output in these years), and 1901–1904 in the report for 1904 (p. 10).

24. Mitchell, *Abstract of British Historical Statistics,* p. 136.

all-time peak only in 1906, with more Bessemer than open hearth steel produced there until 1908, and it is tempting to conclude from this American lag in the adoption of the steelmaking process of the future that British steelmakers were more prescient than their American rivals, turning the usual assessment on its head.[25] The temptation must be resisted, however, for two reasons. First, as will be shown in detail in the next chapter, the use of hindsight to identify technological progressiveness is hazardous, for a technology that becomes profitable to adopt under altered circumstances tomorrow may not be profitable today. Second, as has been emphasized in this chapter, demand as well as supply determines the pattern of output, and demand was largely outside the control of entrepreneurs in British iron and steel. Roughly half of the tonnage of British open hearth products during the 1890's and 1900's were plates and angles for ships.[26] It was the demand for a steel that could bear up under the violent strains to which a ship built of it was subjected and the declining importance in the British market of steel fit only to withstand the compression imposed on rails carrying trains, rather than any special British genius, that made the British industry concentrate on the open hearth process unusually early in its career.

25. W. A. Sinclair makes this point, emphasizing the progressiveness of the open hearth branch of the industry, in "The Growth of the British Steel Industry in the Late Nineteenth Century," *Scottish Journal of Political Economy,* 6 (1969), 33–47. Compare Payne, "Iron and Steel Manufactures," in Aldcroft, ed., *The Development of British Industry,* p. 92. The American statistics for the comparison are from Temin, *Iron and Steel in America,* p. 271, from the American Iron and Steel Association, *Statistics.* Incidentally, the British experience casts some doubt on Temin's argument that in America the basic process was required to make the open hearth profitable (Temin, p. 141); Britain adopted the open hearth earliest, but was the last major steel-producing country to adopt the basic process in the open hearth.

26. In the major shipbuilding areas in Scotland and on the Northeast Coast of England plates and angles were typically two-thirds of open hearth output. The sources for the calculations are, as above, the annual reports of the British Iron Trade Association. In some years the voluntary nature of their survey of the output of products resulted in underreporting, but the error is not large.

The statistical profile of the British iron and steel industry is a composite of elements of demand and supply, and when the elements of demand over which the industry had no control are removed the profile is less clearly one of entrepreneurial failure. Output of iron and steel grew slowly by international standards, to be sure, but this was in part a consequence of slowly growing demand. When branches of the industry could increase the demand for their products, as the steel industry could before 1890, they did. When their markets presented new opportunities, as did the market for shipbuilding steel with the development of the open hearth process, they seized them. The limitation, however, was that demand as a whole was growing slowly, as industrializing nations built up tariffs walls around once-rich markets for British products and as the British economy itself, already industrialized, settled into a more sedate pattern of growth. The stimulus given to the iron and steel industry by the growth of its demand was not large, but this is not in itself evidence of entrepreneurial failure; what remains is to measure the response to the stimulus on the side of supply, and to gauge its adequacy.

4   The "Most Notable Single Instance" of

Entrepreneurial Failure: The Neglect

of the Basic Process

If the picture presented earlier of competition in the iron and steel industry is accurate, one might expect British entrepreneurs to have seized profitable innovations quickly, as the rapid move to the new steel processes suggests they did before 1890. It has been alleged repeatedly, however, that British entrepreneurs were in fact lethargic in adopting innovations compared with their competitors abroad, and after 1890 the allegation appears on some counts convincing. To be sure, the way observers anxious to prove that British entrepreneurs failed have used international comparisons is at times misleading. The British industry, for example, is often condemned for its slow adoption of the by-product coke oven, which was put into general use in the 1880's on the Continent, turning out marketable amounts of tar, benzol, sulphate of ammonia, and other chemical constituents of coal discarded in the older process twenty-five years before even half of the British output came from by-product ovens.[1] Yet the American industry,

1. Burn, *Economic History of Steelmaking,* pp. 206–207: "The new-style coking which proved economical . . . after 1900 [in Britain] could have been so by the middle eighties . . . That the use of the process was so long delayed was due, proximately, to the readiness of blast-furnace owners and managers to accept the results of short and spasmodic experimenting by an authority." The authority was I. L. Bell, an eminent ironmaster, scientifically trained, whom Burn portrays as a conservative force in the industry.

which few would accuse of lethargy, was slower still than the British to move to the by-product process, suggesting that relative factor prices (in this case the price of coal relative to labor) more than relative entrepreneurial vigor determined the extent to which the innovation was adopted.[2] There is one innovation, nonetheless, that Britain was unique in neglecting, namely, the Thomas-Gilchrist or "basic" process of steelmaking. In Duncan Burn's opinion the neglect of the basic process was "the most notable single instance of the slow adoption of new methods."[3]

Was the Neglect of Basic Ores Irrational? The basic process— so called because it requires a chemically basic material, normally limestone, in the furnace walls—removed embrittling phosphorous from pig iron and permitted the ample deposits of phosphoric ores, formerly neglected, to be used in making pig iron for steel. Invented in Britain at the end of the 1870's, the process dominated continental steelmaking by the end of the 1880's, fifty years before it was adopted on a similar scale in Britain. As Clapham remarked, "it is hard to believe that a process employed so extensively in 1925 and 1913 might not have been employed to advantage rather more than it was in 1901 and earlier."[4] Burnham and Hoskins assert that "British ores were adaptable to the Thomas process, but due to prejudice they were not exploited" and that "what was required was the setting up by 1890 of Thomas plants in Lincoln- shire or Frodingham" (where the British phosphoric ores were located).[5] In 1905, J. S. Jeans, Secretary to both the Iron and Steel Institute and the British Iron Trade Association, and manag- ing editor of the *Iron and Coal Trades' Review,* blamed the con- sumers of steel for the oversight: "If the engineers of this country

2. So late as 1918, 75 percent of coke came from by-product ovens in Britain and only 46 percent in America. See Carr and Taplin, *History of the British Steel Industry,* p. 210, and American Iron and Steel Institute, *Statistics, 1918,* p. 63.
3. *Economic History of Steelmaking,* p. 182.
4. *Economic History of Modern Britain,* vol. III, p. 148.
5. *Iron and Steel in Britain,* pp. 120, 180.

were as ready to accept basic steel as those on the Continent, it might be possible to greatly diminish our dependence on imported ores. I am aware of numerous ore deposits in this country that would, in that event, be much more fully utilized than they are at present."[6] In 1890, P. C. Gilchrist, one of the inventors of the process, blamed the manufacturers: Germany's much larger production of basic steel (three times larger than Britain's at the time) was "due not to their having better ore for the process, but to the fact that they have as officers in their works men who are alike practical and theoretical and who are able to avail themselves quickly of any new process."[7] A substantial body of contemporary and historical opinion, in short, has agreed with Burn that the neglect of the basic process was a serious failure. The question is to what extent their opinions are justified.

Burn's careful discussion of the issue is the natural place to begin.[8] The demonstration of an entrepreneurial failure in basic steelmaking is the keystone of his book and has been used repeatedly in general economic histories as an example of failures in the British iron and steel industry and in the economy as a whole in the late nineteenth century.[9] Burn was attempting to explain

6. *The Iron Trade of Great Britain* (London: Methuen, 1906), p. 15. Immediately after this assertion Jeans remarks that "acid-steel . . . [is] specially suited and applied to the building of ships, and the manufacture of tyres, axles, wheels, and other important railway requirements" (p. 16). If acid steel was "specially suited" to shipbuilding, it is not surprising that basic steel was neglected, because a large part of open hearth output went ultimately to shipbuilding materials. In this case it would be difficult to demonstrate an irrational neglect of basic steel on the part of consumers.

7. *Proceedings of the Cleveland Institution of Engineers* (Middlesbrough), 1890–1891, p. 131.

8. *Economic History of Steelmaking,* chap. IX, "New Ores for Steel; or, the Advantages of Location," pp. 151–182.

9. S. Pollard and D. W. Crossley, in *Wealth of Britain,* p. 227 write: "British entrepreneurs seemed to be unable to match the real cost reductions of other countries . . . In what is perhaps the outstanding example, the iron and steel industry, they neglected to work the low-grade home ore deposits which should have provided them with cheaper pig iron than they were then able to obtain from abroad." Compare W. Ashworth, *Economic History of Modern Britain, 1870 to 1939,* p. 87; E. J. Hobs-

why British steel prices became higher than foreign prices around 1900. One contributing cause, he said, was that the supply curve of the traditional acid steel of Britain was rising because diminishing returns to ore extraction were not offset by transport innovations, as they were in the United States. The central cause, however, was that the supply curves of foreign producers of steel, who unlike the British steelmakers had moved rapidly from acid to basic steel in the 1880's (Germany) or in the 1890's (Belgium and America), were falling: "The absence of improvements in transport was a far less impressive source of high costs than the comparative neglect of the cheap phosphoric ores which Sidney Thomas had made available for steel. This was, indeed, the most amazing feature of British steelmaking."[10] Burn conceded that there were good reasons for the comparative neglect of basic steelmaking in the old centers of the trade such as North Yorkshire. The ores of North Yorkshire, in fact, had too *little* phosphorous for the effective use of the basic process in its early forms. For the neglect of the phosphoric ores of Lincolnshire and Northamptonshire, however, he could see no adequate explanation. This, then, was his fundamental criticism of the industry, namely, that Lincolnshire and Northamptonshire should have been centers of the pig iron industry. In failing to relocate the industry, the managers were neglecting a cheap source of pig iron for wrought iron as much as for steel. Put this way, it is clear that Burn's argument is only secondarily concerned with "the weakness of the basic steel industry in Britain";[11] it is primarily concerned with the extroardinary error of the manufacturers of pig iron, wrought iron, and steel in failing to exploit the Lincolnshire and Northamptonshire ores.[12]

---

bawm, *Industry and Empire,* p. 158; P. Mathias, *First Industrial Nation,* p. 411. But contrast R. S. Sayers *History of Economic Change in England, 1880–1939,* p. 82.
   10. Burn, *Economic History of Steelmaking,* p. 182.
   11. Ibid., p. 182.
   12. Cf. ibid., p. 335.

Burn conceded, first, that the only cheap native phosphoric ores perfectly suitable for steelmaking were the lower lias ores of North Lincolnshire, even these being "rather lean, rather variable in quality, and not very phosphoric";[13] and, second, that these ores were so far from consuming centers that "what they gained on the ore . . . they might lose on the trains."[14] These two concessions can be developed into a strong case against Burn's allegation of irrational neglect. Burn's assertion is that basic pig iron could be produced for very much less in North Lincolnshire than anywhere else in Britain, but transport costs to markets, the high cost of moving production to Lincolnshire, and most important, sheer irrationality kept the production of pig iron there low. The argument can be described diagrammatically, as shown in Figure 2. In consequence of the cost of transportation and the irrational neglect of the North Lincolnshire ores, the supply curve of Lincolnshire to the market is raised from S_L to S_L', by the length of the segment $C_L'\ P_L'$; the market equilibrium, which is determined by the horizontal sum of the Lincolnshire supply curve and the Cleveland supply curve (S_C' standing for all other sources of supply: the Cleveland district of North Yorkshire produced a third of all British pig iron), is at the point E' rather than at the point of equilibrium in the absence of transport costs or irrational neglect, E; the quantity of production in Cleveland (Q_C') is larger and the quantity in Lincolnshire smaller (Q_L') than they would be in the alternative circumstances The difference between the price of Lincolnshire pig iron at the point of consumption, taken to be Cleveland, and the cost of production of Lincolnshire pig iron is the segment $C_L'P_L'$: part of it is the mere cost of transport; the rest is attributable to irrationality, and can be isolated by comparing the market price, P_L', with the true supply price, C_L', after allowing for transport costs.

One sort of irrationality can be measured by comparing the supply prices including rents and another sort by comparing supply prices excluding rents. "Static irrationality," as one might call it,

13. Ibid., p. 168.
14. Ibid., p. 181.

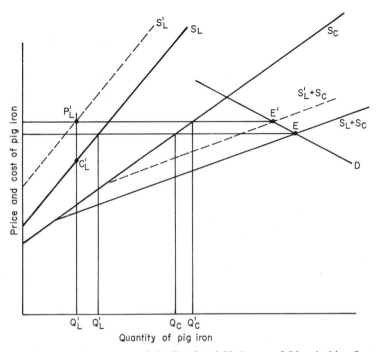

Fig. 2. The Consequences of the Irrational Underuse of Lincolnshire Ores and of the High Cost of Transporting Lincolnshire Pig Iron to Centers of Consumption

is failing to buy at the lowest available price and to sell at the highest, and it is necessary to test for its existence in order to validate the use of any model that assumes arbitrage in the market, as this one does. If Lincolnshire pig iron sold for less than Cleveland pig iron of the same grade in the same market, continued exchanges of Cleveland pig iron would demonstrate the static irrationality of consumers in not buying iron at the lowest price and the static irrationality of sellers in not selling at the highest price. The relevant statistic for this test is the prevailing market price, inclusive of rents: if the prices differ substantially there is substantial static irrationality. "Dynamic irrationality" is failing to

61

expand production when one's activity is earning pure profit or, in other words, Marshallian quasirents. If the costs excluding rents of Lincolnshire pig iron were substantially less than those of Cleveland pig iron, Lincolnshire pig iron was earning a differential rent and its output should have been expanded relative to Cleveland pig iron. Interpreting the diagram this way, if C_L', the cost of Lincolnshire pig iron excluding rent, was substantially below P_L', which is equal to the cost of Cleveland pig iron, each Lincolnshire manufacturer of pig iron would have an incentive to increase his output, on the expectation that he could sell the additional output at more than the additional cost of producing it. Dynamic irrationality, then, would result in a persistent difference between P_L' and C_L', and the relevant statistic for the test is the cost difference, excluding rent, between Lincolnshire and Cleveland pig iron.[15] Burn believed that the cost of Lincoln pig iron was very far below the cost of Cleveland pig iron.[16] It can be shown, however, that C_L' in the diagram was in fact in the neighborhood of P_L', whether the cost curves are defined to include or exclude rents. That is, the industry *was* at the optimal equilibrium E.

Consider first the test for static irrationality, comparing the market prices of Lincolnshire and Cleveland pig iron. There is unfortunately little direct information on the price of Lincolnshire pig iron before 1914. The only price available is the price of

15. There is a notable parallel between these two tests of rationality and tests of the profitability of American slavery. The static test, like that of A. H. Conrad and J. R. Meyer in "The Economics of Slavery in the Ante Bellum South," *Journal of Political Economy*, 66 (1958), 95–130, focuses on how well the market was functioning. The dynamic test, like that of Y. Yasuba in "The Profitability and Viability of Plantation Slavery in the United States," *Economic Studies Quarterly*, 12 (1961), 60–67, reprinted in Robert W. Fogel and Stanley L. Engerman, eds., *The Reinterpretation of American Economic History* (New York: Harper & Row, 1971), focuses on how much the industry could profitably have been expanded.

16. "Lincolnshire pig could be sold profitably for thirty shillings a ton in the early nineteen-hundreds" (Burn, *Economic History of Steelmaking*, p. 168). The price of pig iron at that time was forty or fifty shillings a ton.

Lincolnshire number III foundry pig iron in Lancashire.[17] The railway rate from Lincolnshire to Lancashire was around 9 shillings per ton of pig iron.[18] Subtracting this rate from the price in Lancashire (in order to estimate the price in Lincolnshire) and comparing the result with the price of Cleveland number III foundry pig iron in Cleveland, one finds that from 1897 to 1913 the price of Cleveland iron averaged about 3.3 shillings more per ton than the price of Lincolnshire iron. This is well within the range of error: Cleveland iron, for example, may have been consistently 3 shillings better in quality than Lincolnshire iron or the railway rate may have been less. More important, it is well within the cost of transporting iron from Lincolnshire to Cleveland, which would have been about as expensive as transporting it to Lancashire. Only if the difference in price exceeded the transport cost would there be evidence of static irrationality. As it is, the cost of transport in Britain would have to have been a third to a half of what it was for the observed price difference to indicate static irrationality.

It is also possible to estimate the difference in the market prices of the two pig irons from the difference in the market prices of their raw materials. The major difference between Cleveland and Lincolnshire as sites for basic pig iron production was that Cleveland was relatively close to high-quality coking coal and Lincolnshire was relatively close to cheaply quarried iron ore. The closest coking coal to North Lincolnshire was around Leeds in South Yorkshire, a distance of about fifty miles, whereas the North Yorkshire ore fields were only about twenty miles from the high-quality coking coal of Durham. The price of coke at the ovens

17. It is available in the weekly price lists of the *Iron and Coal Trades' Review* from the 1890's onward.

18. This is the figure for class C goods (which includes pig iron) from Lincolnshire to Liverpool quoted in J. S. Jeans's evidence to the Royal Commission on Canals and Waterways for 1907 (in S.P., p. 87). Railway rates before the war were very stable, by law. Jeans remarks on p. 86 that "the railway rates of today are, generally speaking, much the same as those of a quarter of a century ago."

(close to the mines) was roughly the same in Durham and York-shire. The difference in coke cost between Lincolnshire and Cleve-land, then, would be simply the extra freight on the coke multiplied by the amount of coke required per ton of pig iron. The difference in ore costs can be calculated from the at-mine value of the Cleve-land and Lincolnshire ores given annually in *Mineral Statistics of the United Kingdom*. The physical requirements of coke and ore can be calculated from the same source, although only the Lincoln-shire requirements are used in the comparison here because the North Yorkshire pig iron returns do not distinguish hematite (acid) from ordinary pig iron production. This procedure will bias the calculation towards finding Lincolnshire costs lower relative to Cleveland costs than they in fact were, because Cleveland makers would have used less ore and more coke, if anything, in accord with the lower relative cost of coke in Cleveland than in Lincoln-shire, thereby achieving lower costs relative to Lincolnshire than those implied here. The one difficulty is the variation in contempo-rary estimates of the relative cost of coke transport in Lincolnshire and Cleveland. J. S. Jeans testified to the Tariff Commission of 1904 that the cost of transporting coke to the blast furnaces was 2.5 shillings in Cleveland and 4.5 shillings in Lincolnshire, a difference of 2 shillings.[19] The testimony of Jeans and others to the Royal Commission on Canals and Waterways in 1907, however, suggests that 2 shillings is a low estimate, 3 or even 4 shillings being perhaps more accurate. Table 3, therefore, calculates for 1890, 1899, and 1909 the difference in materials costs in Cleve-land and Lincolnshire assuming 2-, 3-, and 4-shilling differences in transport costs.

The largest difference in total costs is 4 shillings in 1909 assuming

19. *Report of the Tariff Commission,* vol. I: *The Iron and Steel Trades* (London: Tariff Commission, 1904), par. 1127.

TABLE 3. Difference in Combined Ore and Coke Costs in Making Basic Pig Iron in Cleveland and Lincolnshire, for a Range of Values of the Transport Cost of Coke in Lincolnshire, 1890, 1899, 1909.
(in shillings per ton of iron)

| Year | Amount, for coke transportation costs (in shillings per ton of coke) of | | |
	2s	3s	4s
1890	2.4	1.1	−0.17
1899	2.5	0.88	− .71
1909	4.0	2.5	1.1

Source: All the data for the calculation are from *Mineral Statistics of the United Kingdom* for the years given, except for the assumed coke transport costs. The procedure was to multiply the observed Lincolnshire requirements of coke and ore per ton of pig iron by the difference in ore costs at the mines and coke costs (including transport) at the blast furnaces. The data on ore costs at the mines for "Lincolnshire" are in fact for Lincolnshire and Leicestershire together, as these are not reported separately in *Mineral Statistics;* the methods of mining (quarrying in this case) and the quality of ore were similar in the two counties.

a transport cost difference of only 2 shillings. As in the previous test, the difference in estimated market prices in Lincolnshire and Cleveland is well below the cost of transporting pig iron between the two places and, therefore, indicates no static irrationality.

The test for dynamic irrationality can be carried out in much the same way, balancing the higher transport cost on coke against the lower cost of ore, except that now the substantial element of rent in the cost of ore must be eliminated. Since the issue is one of an allegedly *larger* incentive to expand the output of pig iron using Lincolnshire ore relative to that using Cleveland ore, the "dynamic" test should be undertaken excluding rent in both locales: a nationwide incentive to expand output in all regions may exist without differential incentive among regions. I. L. Bell, writing in 1886, asserted that royalties on ore were at times a shilling a ton in Lincolnshire compared with half that in Cleveland because

Lincolnshire ore was quarried and Cleveland ore mined.[20] If the resource cost of quarrying Lincolnshire ore were below the cost of mining Cleveland ore, then the market price of Lincolnshire ore, given the earlier evidence from the fact of adequate arbitrage that Lincolnshire and Cleveland pig iron were substitutes in consumption, would tend to rise until the money costs of the two varieties of pig iron were equalized by rents. It is for this reason that comparisons of the profitability of techniques (to establish the rationality of the entrepreneurs involved), if the techniques are employed in the same market and if the comparisons use market prices of inputs including rents, are biased towards finding no difference at all. The question is, then, were the resource costs of Lincolnshire and Cleveland pig iron so different that the rents earned in Lincolnshire were in fact substantially higher than those earned in Cleveland?

The testimony of Jeans to the Tariff Commission of 1904 can be used for this as for the previous calculation. He gives the costs of transporting a ton of coke to the blast furnace as about 4.5 shillings in Lincolnshire and 2.5 shillings in Cleveland, as has been said, and the cost of extraction of ore (not the market price, which would include rents) as about 2.0 shillings per ton in Lincolnshire and 2.5 shillings per ton in Cleveland.[21] If pig iron made in Cleveland had required 3.3 tons of ore and 1.5 tons of coke per ton of pig iron, as it did at the time in Lincolnshire, it would have cost 1.4 shillings per ton *less* than Lincolnshire pig iron.[22] The difference in costs is slight, and no weight should be put on the find-

20. I. L. Bell, *The Iron Trade of the United Kingdom* (London: British Iron Trade Association, 1886), p. 75.

21. *Report of the Tariff Commission,* as cited above, pars. 1131 and 1127.

22. The calculation is $(2.5 - 2.0)(3.3) + (2.5 - 4.5)(1.5) = -1.4$ shillings, Jeans gives ranges for the two types of costs in each district. Assuming that the factor proportions are exact, these ranges (plus or minus sixpence on each cost element) imply a range in the final cost difference from -6.15 shillings to $+3.45$ shillings.

ing that production in Cleveland was cheaper. What is impotrant is that the costs were virtually equal. In terms of the previous diagram, Figure 2, the costs of Cleveland and Lincolnshire pig iron were virtually the same, the industry was in equilibrium in the neighborhood of point *E,* and nothing would have been gained by an increase in the relative output of Lincolnshire pig iron.

The result is robust. It is not entirely clear that Jeans removed royalties from his estimates of "extraction costs," although he does decompose the costs into the average product of labor and its wage, which suggests that only labor costs, not royalties as well, were included. Supposing he did not, however, does not greatly change the result. In 1893 J. D. Kendall gave detailed estimates of the extraction costs of Lincolnshire ore and the royalty payments on Cleveland ore.[23] Applying these estimates to 1904, the total advantage of Lincolnshire becomes +0.06 shillings per ton of pig iron, at a time when pig iron of this grade was selling at 40 or 50 shillings per ton.[24] The relative neglect of the Lincolnshire ores, in short, appears to have been a rational adjustment to the balance of locational advantage.

23. *The Iron Ores of Great Britain and Ireland* (London: 1893), pp. 374–379.

24. The procedure is to subtract the 1893 estimate of Cleveland royalty payments (0.50s) from the 1904 Cleveland "extraction cost" (2.50s) and to replace the 1904 Lincolnshire "extraction cost" (2.00s) with the 1893 estimate (1.08s). The ore term, then, becomes (2.5 − Cleveland 1893 royalty) − Lincolnshire 1893 extraction costs. Because the price of domestic ore fell from 5.05 shillings a ton to 4.55 shillings between 1893 and 1904, using 1893 data may slightly understate the cost advantage of Lincolnshire, the subtracted data for 1893 on Cleveland royalties and Lincolnshire extraction costs being perhaps too high for 1904. The rent on Lincolnshire ore was so large a part of the market price of the ore that the extraction costs may have risen or fallen even though the price fell 10 percent. The rising price of labor between these dates would suggest that the extraction costs, in fact, rose. Even if the Lincolnshire extraction costs remained the same and all the Cleveland 1893 royalties on ore (amounting to 0.50s) were eliminated by the 0.50s fall in the average ore price, the advantage of Lincolnshire pig iron would be increased to only 1.7 shillings, well within the range of an insignificant difference.

The Shift to Basic Steel. Explaining the neglect of basic ore, however, does not necessarily explain the neglect of basic steel: the exploitation of the Lincolnshire ores in the making of pig iron and the adoption of the basic process in steelmaking were not necessarily connected either technically or historically. This being so, the question whether British basic pig iron production was irrationally retarded and the question whether, given the location of pig iron production, basic steel production was irrationally retarded are more separate than the usual treatment of the two together would suggest. The key technical fact is that the basic open hearth process could use a wider range of ores than the acid open hearth or Bessemer processes or the basic Bessemer process. The Lincolnshire and Northamptonshire ores were, to be sure, made available for steelmaking by the basic open hearth process, but so too were all the other native ores used formerly only for puddled or cast iron production. Moreover, if the Lincolnshire ores were suitable for basic open hearth steelmaking, they were also suitable for puddled and cast iron production. In short, not all basic steelmakers were users of the Lincolnshire ores and not all those who neglected the Lincolnshire ores were steelmakers.

The technical disconnectedness of the Lincolnshire ores and basic steelmaking shows clearly in the historical record. During the 1890's the output of basic steel from both the open hearth and the Bessemer converter averaged 0.56 million tons per year, or 16 percent of the total annual output of steel; during the 1920's it was 4.7 million tons, 74 percent of the total. On the most generous assumptions, however, only 36 percent of the absolute increase of 4.1 million tons per year in basic steel tonnage between the two decades can be attributed to the increase in the use of the ores of Lincolnshire and of similar counties. The average annual output of ore in all the counties whose deposits could be considered to have been neglected in the late nineteenth century—the East Midlands counties of Leicestershire, Northamptonshire, Rutland, and Oxfordshire as well as Lincolnshire itself—increased from 3.4 million tons in the 1890's to 5.9 million tons in the 1920's. Estimating the iron content of these ores at 30 percent (a high estimate, particularly after the ore fields came to be exploited more intensively: in the 1920's 27 percent is more realistic) and assuming that none of the ores were used to produce cast or wrought iron and that there was no wastage of the iron content, the increased output of phosphoric ores could account for an increase in the annual output of pig iron of 0.74 million tons. Open hearth basic steel was made with scrap in addition to the pig iron and, in contrast to the 1890's, when 74 percent of basic steel was made in Bessemer converters, in the 1920's virtually all basic steel came from the open hearth. The ratio of basic steel output to the pig iron used to produce it in the 1920's, therefore, will serve as an upper bound on the ratio in previous years; the highest value it attained in the seven years after the statistic first became available (1920) was 2.0, in 1923. At most, then, the 0.74-million-ton increase in basic pig iron made with East Midlands ore can account for 1.48 million tons of the total increase of 4.1 million tons in the annual output of basic steel between the 1890's and the 1920's. In other words, at the very least nearly two-thirds of the increase in the output of basic steel was achieved without recourse to the "neglected" ores.[25] The adoption of the basic process in Britain cannot be identified with the adoption of native phosphoric ores in the East Midlands.[26]

25. The statistics of pig iron and ore output for 1890–1899 and 1920–1929 used in this paragraph are from *Mineral Statistics of the United Kingdom* and the British Iron and Steel Federation, *Statistical Yearbooks* (London, annually after 1915, under varying titles); Mitchell, *Abstract of British Historical Statistics* presents them in a convenient form at pp. 129–132. The iron content of Jurassic (East Midlands) ores other than those of Cleveland are given in the B.I.S.F. *Yearbooks* for the 1920's (as are the scrap ratios) and in *Mineral Statistics* for the 1890's.

26. Contrast Burn's comment on "How the War Affected British Competitive Strength" (chap. 14 in his *Economic History of Steelmaking*): "The industry and its customers were forced at long last to accept the basic process without irrational reservations, and the East Midlands ore

69

The emphasis on the slow move to the basic ores has obscured the fact that independent of the use of the ores there was a rapid move in the early 1900's to basic steelmaking: from 1900 to 1913 basic open hearth steel output in Britain grew at 15.6 percent per year, while acid open hearth output grew at only 2.2 percent per year and all steel at 3.2 percent per year. The lethargy of British steelmakers in adopting the basic process after 1900 has been exaggerated. To what extent the charge of lethargy is justified earlier, in the 1890's, is difficult to say, because it is more difficult for the basic steel industry than for the Lincolnshire pig iron industry to perform the appropriate tests of rationality. What can be done, however, is to make it plausible that the ultimate adoption of the basic process had an economic rational independent of entrepreneurial awakening in the 1900's from an earlier slumber. One explanation, on which Burn puts some emphasis, is that in the 1900's growing demand pushed the industry into a region of inelasticity in the supply curve of acid pig iron, and with more expensive acid pig iron acid steelmaking became less profitable. An alternative is that for technological reasons in the 1900's the supply curve of basic steel moved out. Although hardly conclusive, the weight of evidence favors this latter, technological, explanation.

The rise of the basic process in the 1900's was the rise of the basic open hearth furnace, not the basic Bessemer converter: there was little tendency for the ratio of acid to basic steel produced from the Bessemer converter to change during the 1890's and 1900's. From 1895–1899 to 1909–1913 the share of basic Bessemer in all Bessemer steel produced rose only slightly, from 28 percent to 35 percent, while the share of basic open hearth in all open hearth steel rose from 8 percent to 36 percent. Apparently, then, it was neither a rise in the cost of acid relative to basic pig iron nor a new realization that the basic process was cheaper, both of which

---

resources were systematically explored" (p. 350). The identification of the two phenomena is made throughout the book. Incidentally, it is unnatural to speak of the war "forcing" the adoption of the basic process: on its eve 37 percent of open hearth steel output was basic.

could be expected to have changed the share of basic steel in both the open hearth furnace and the Bessemer converter more or less symmetrically, that impelled British steelmakers to produce more basic steel. The balance of advantage must have shifted to an exceptional degree in favor of the basic open hearth alone.

The particular innovation that appears to have been responsible for the shift in relative advantage is the Talbot tilting furnace. The open hearth had the disadvantage, avoided in Bessemer practice by charging hot metal, that time and fuel were wasted in charging the furnace, melting the charge, and cooling the furnace for the next charge. In the early 1890's Benjamin Talbot, an Englishman working at the Pencoyd Steelworks in Pennsylvania, experimented with keeping the open hearth hot for long periods of time, adding hot metal in small amounts, and removing the finished part of the charge by tilting the furnace slightly.[27] When introduced commercially in 1900, the Talbot tilting furnace was hailed as "the greatest advance that had been made in the manufacture of steel for some years."[28] By 1910 there were ten Talbot furnaces operating in Great Britain.[29] Although useful in acid steelmaking, the new furnace was especially useful in basic steelmaking. The basic process produced great amounts of slag, which increased the consumption of pig iron by absorbing iron, increased the consumption of coal and time by shielding the metal from the fire, and increased the corrosion of the furnace bottom by reacting with it when the furnace was emptied. In the Talbot process, steelmakers could keep the blanket of slag thin by tipping it out of the furnace, reduce the periodic strain on slag-handling equipment by delivering

27. Talbot acknowledged that his experiments were inspired by the success of a tilting furnace designed by the Americans Campbell and Wellman at Pencoyd.

28. G. J. Snelus, vice-president of the Iron and Steel Institute, in *Journal of the Iron and Steel Institute,* 1900, pt. I, p. 62.

29. There were many in the United States, but only two in Germany, reflecting the balance of advantage in the handling of slag mentioned below. An article by O. Peterson in *Stahl und Eisen,* which gives the distribution of Talbot furnaces, is summarized in the *Iron and Coal Trades' Review,* 1910, p. 210. Talbot furnaces were larger than other types.

71

the slag in small amounts at frequent intervals, and could prevent the slag from corroding the bottom by seldom completely emptying the furnace. The Talbot process, in other words, reduced the cost differential between the acid and the basic open hearth processes. The effect of reducing the costs associated with the slag was particularly important in Britain because by international standards the slag burden there was unusually large. Scrap, which is free from impurities and therefore contributes little to the generation of slag in the furnace, was expensive in Britain and in consequence the ratio of scrap to pig iron in the furnace charge was small and the amount of slag generated was large.[30] The Talbot process had relatively more to offer British steelmakers, and when it was invented they moved rapidly to the basic process.

The rapid shift to basic steelmaking in the decade before World War I, then, appears to have a rational explanation. It can be explained by developments in the technology of basic steelmaking rather than by the fading of irrational prejudices against the process. In any case, there *was* a rapid shift. The period of irrational neglect of basic steel, which, even setting aside the evidence that the basic process was unprofitable in Britain before the coming of the Talbot furnace, was at the most the ten years or so in the 1890's when America shifted to basic steel while Britain did not, is a brief episode on which to base an unfavorable assessment of British entrepreneurial performance from 1870 to 1913. And if the assessment is based on the neglect of the Lincolnshire ores, it is mistaken: they were used as much as their cost and location warranted. In short, the charge of an important irrational neglect of basic ores and basic steel, which has long been a central theme in the case for entrepreneurial failure in British iron and steel, is poorly founded.

30. On the Continent basic steelmakers used a charge that was 80 percent scrap and in America 50 percent scrap, while British basic steelmakers used only 20 percent scrap. The market price of scrap appears to have been 15 percent higher relative to the price of pig iron in Britain than in America.

5   Productivity Change, 1870–1913

The iron and steel industry was competitive, was limited by the growth of its demand, and was justified in its neglect of the basic process: the argument so far has established the good character of the accused, has offered evidence in extenuation of its apparent lapses, and has cleared it of the major charge. In the court of historical opinion, however, the industry has been indicted on a great many counts. It would be tedious to examine each separately, and inconclusive as well, for bad entrepreneurship need not always have taken the simple form of the neglect of easily identifiable innovations. Some broader testimony on entrepreneurial performance is required. It will be provided now by measuring the industry's productivity change.

The relevance of measures of productivity change is that they isolate the effect on output of entrepreneurial performance by removing the effect of the industry's inputs. The calculation is possible because marginal productivity theory, which is especially simple in the competitive milieu of the iron and steel industry, provides a plausible estimate of how inputs affect output. That is, those influences on output that can be measured can be removed, leaving a residual change in output attributable to unmeasurable influences, among them entrepreneurial performance. Since no productivity measure, least of all one based on the sparse and fragile information available on the iron and steel industry in the late nineteenth century, can bring all the inputs to the calculation, this residual will include measurement errors as well. The argument of this and the following chapters, therefore, attempts to stay within wide bounds of error. The major remaining difficulty is that advances in technological knowledge as well as changes in entrepreneurial skill in adopting the knowledge are reflected in the measures. To isolate entrepreneurial skill, in other words, the effect of worldwide accelerations or decelerations in technology must be removed from the measures of productivity in Britain, and to accomplish this the subsequent chapters compare them with measures of productivity in America.

The Material-Intensity and Capital-Lightness of the Industry. Before turning to the central task it is necessary to make a point about productivity measurement and to stress the relevance to that measurement of the iron and steel industry's peculiar structure of inputs. "Productivity" is customarily defined as output per man or output per composite unit of men and machines, setting aside inputs of materials from other industries. Although this definition is appropriate for measuring productivity in the nation as a whole, it is not for measuring it in one industry alone, whatever the end in view. It is inappropriate if the measure is meant to reflect the increased national income generated by technological change or improved efficiency in the industry, for these events release for alternative employment the labor and capital embodied in materials used by the industry as well as the labor and capital used directly. And it is also inappropriate if the measure is meant to reflect the responsiveness of entrepreneurs to market pressures to minimize costs, for these pressures induce entrepreneurs to save materials as well as labor and capital directly employed in the industry. Measures of productivity change for single industries should include material inputs.

This observation has special significance for the iron and steel industry. About two-thirds of the total costs of the British iron and steel industry at the time of the Census of Production of 1907 were costs of materials purchased from other industries, such as coal and ore: of the industry's total costs of £90 million, a mere £30 million were direct expenditures on labor and capital.[1] Among

1. See Appendix C for the analysis of the input and output structure of the industry in 1907 from which these figures are derived.

the other industries described in the Census, only the Metal Trades other than Iron and Steel (largely the working of gold and silver) and the Leather Trades had higher shares of material costs, and the average for all trades was less than half of total costs. If the iron and steel industry is broken into its component parts, as it is below for purposes of measuring productivity, the picture is similar: ore and coke costs together were three-quarters of the total costs of pig iron, and pig iron was in turn half of the total costs of rolled products of open hearth steel and two-thirds those of Bessemer steel. Productivity change is measured by subtracting a weighted index of inputs from an index of output. The appropriate weights are the shares of each input in total costs, because these shares, for a competitive industry in equilibrium with a production function that exhibits constant returns to scale, are equal to the elasticities of output with respect to changes in the quantities of each of the inputs. Material inputs are easy to measure and have high shares in the costs of iron and steel. Productivity change, therefore, is relatively easy to measure in iron and steel because the measurement is less dependent on information on inputs of labor and capital, which is sparse.

It is most sparse on the input of capital. Fortunately, capital had an especially low share in the total costs of iron and steel. Its low share and relative unimportance for measurements of productivity change is not merely a consequence of the low share of value-added (the returns to labor and capital together, that is, to inputs other than those purchased from other industries), for capital in iron and steel accounted for an unusually low share of value-added itself. The Census of Production of 1907 gives the number of employees in the industry. It does not give the average wage, but for 1906 the Board of Trade undertook an *Enquiry into the Earnings and Hours of Labour* from which it can be estimated.[2]

2. Pt. VI. *Metal Engineering and Shipbuilding Trades in 1906,* S.P., 1911, vol. 88. The *Enquiry* sample of earnings was voluntary and had, unfortunately, a poor response in iron and steel (only about a quarter of actual iron and steel employment was reported). There are clear geographical biases in the sample but they appear to offset one another: ex-

The returns to capital implied by subtracting the wage and salary bill from total value added reported in the Census is about £7.7 million, only 25 percent of the total and only 8.5 percent of total costs. This was low by national standards. Other industries, as we have seen, had higher shares of value-added in total costs; and the shares of capital in their value-added, which for the nation as a whole is the share of capital in home-produced national income, was about 44 percent, well above the 25 percent in iron and steel.[3] It is perhaps the spectacular visibility of the capital stock in iron and steel—the massive accretions of bricks and machinery in furnaces, converters, and rolling mills—that has led observers to take the industry as the model of capital-intensive production. Quite the contrary, however, the iron and steel industry was in fact unusually capital-light.

The measurement of the capital input, then, can be treated in what follows rather casually, for it will have little impact on the measurement of total factor productivity. There is another, admittedly more speculative, inference to be drawn from the low share. The share of capital is calculated as a residual and includes therefore the share of entrepreneurs. Consequently the share of

---

periments with reweightings by regional outputs yielded almost identical averages as the sample averages given on pp. 20 and 30 of the *Enquiry* report, namely, about £80 per year per wage earner. This was adjusted by A. L. Bowley's index of wages (given in Mitchell, *Abstract of British Historical Statistics,* p. 344) to 1907, to give £84.5. The 5.5 percent increase in Bowley's index is in rough agreement with the 6.2 percent increase in the gross value of the industry's output from 1906 to 1907 (estimated in the Census, p. 105). Taking salary earners at £100 per year on average, the total wage and salary bill was about £22.3 million. The residual return to capital (£7.7 million) agrees roughly with the Census estimate (p. 67) that the value of the capital stock was about 2.25 times value added in iron, steel, and engineering; with a typical return on industrial capital of about 11 percent (see E. H. Phelps-Brown and B. Weber, "Accumulation, Productivity and Distribution in the British Economy, 1870–1938," *Economic Journal,* 63 [1953], 263–288, esp. Fig. 3) the return would be about £7 million. If the wage bill were biased upward by as much as 15 percent, the true share of capital in value added would be about one-third, still low by national standards.

3. See Appendix A for the estimate of the national share of capital.

entrepreneurs in the costs of British iron and steel must have been well below 8.5 percent. To the extent that it is reasonable to speak of entrepreneurship as an input like any other, then, the elasticity of output with respect to the entrepreneurial input would have been small.[4] Suppose, purely for illustration, that the age, education, and other characteristics of British entrepreneurs suggested that over the years from 1870 to 1913 the quality of entrepreneurship corrected for quality fell as much as 100 percent below what it could and should have been and that its share was as much as 5 percent of costs. By 1913, therefore, output would have been only 5 percent below what it would have been had entrepreneurs maintained their earlier vigor. In other words, because of its low share in cost even a large deterioration in the entrepreneurial input would have had a relatively small depressing effect on output. Plausible measures of the entrepreneurial input are so difficult to achieve that this test cannot be performed, but the low share does nonetheless put narrow limits on the impact of an entrepreneurial failure. To put the argument the other way around, if the measures of productivity do in fact register a substantial entrepreneurial failure, the deterioration in the entrepreneurial input in Britain must have been very large indeed.

Pig Iron. It is convenient for measuring productivity to treat the pig iron branch of the industry separately from the others because the material-intensity of its processes and the volume of its statistical records make possible especially simple and complete measures. The simplest measure, developed here in detail because it serves later as the basis for international comparisons, is the output of pig iron per ton of coke used, which was widely accepted by contemporary ironmasters as an index of blast furnace performance; its inverse was given a name, the "coke rate." In looking back on the history of the industry some observers have chosen

4. For a fuller development of this line of reasoning, applied to American farming, see Yair Mundlak, "Empirical Production Function Free of Management Bias," *Journal of Farm Economics,* 43 (1961), 44–56.

Economic Maturity and Entrepreneurial Decline

output per man as the measure of productivity, but in view of the low share of labor in total cost this procedure is not appropriate. The costs of labor were only 10 percent of the total in pig iron, while the costs of coke were 30 percent: if all inputs must be represented by one, the one with the largest share should be chosen, unless its productivity is known to be unchanging. Ore's share in cost was 45 percent, but ore productivity is disqualified from serving as an index because throughout the late nineteenth century ore was fixed in rigid proportion to output, determined by the iron content of the ore. That is, productivity change was biased towards coke, labor, and capital and away from ore, making output per ton of ore a bad indicator of the productivity change affecting the use of other inputs.[5]

The average product of coke, then, is likely to be a reasonable measure of productivity in pig iron. As can be seen in Figure 3, it rose quickly from 1870 to a peak in 1885, but thereafter

5. These plausible assertions can be given a precise form as follows. Following Solow and others, postulate a production function with constant returns to scale and biased productivity change affecting the quantity required of coke, capital, and labor inputs (C, K, and L) for a given output (Q) by the proportions γ, κ, λ, which vary over time (usually upwards because productivity usually increases):
$$Q = F(\gamma C, O, \kappa K, \lambda L).$$
Notice that ore requirements are assumed not to be subject to productivity improvements. Denoting proportional changes in a variable by asterisks and shares of inputs in cost by S_c, S_o, and so forth, the usual manipulations show that, assuming competitive equilibrium $S_c\gamma^* + S_K\kappa^* + S_L\lambda^*$ is an index of the proportional change in output over the period in question that cannot be explained by changes in the quantities of inputs. The question is how well a simple average product of coke approximates this ideal. The proportional change in the average product over some period of time is

$$Q^* - C^* = \frac{1}{S_c} [S_c\gamma^* + S_K\kappa^* + S_L\lambda^*]$$

$$- \frac{1}{S_c} [S_K (Q^* - K^*) + S_L (Q^* - L^*) + S_o (Q^* - O^*)].$$

If S_K and S_L are small, the second bracketed term will tend to be small (note that historically $Q^* - O^*$ is zero) and the average product will be a good approximation to the complete measure.

78

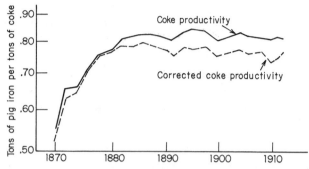

Fig. 3. The Average Product of Coke, 1870–1912

stagnated (ignore for the moment the curve marked "Corrected coke productivity").[6] The stagnation is striking, but two possible explanations of it need to be examined before it can be accepted as a reasonably certain indication that there was no improvement in the technology of the pig iron industry after the mid-1880's.

One possibility is that the quality of iron ore used in the blast furnaces deteriorated. Ore has many dimensions of quality, but the most relevant here is its iron content, because the higher the iron content the less coke is needed to burn the natural oxygen out of the ore: in the limit, of course, an "ore" composed of pure iron would require no coke at all. The iron content of ore varied a great deal from district to district in Britain, from under 30 percent in the blast furnaces of Lincolnshire, which used domestic quarried ores, to over 50 percent in South Wales, which used rich ores imported from northern Spain, yet so stable was the geographical distribution of the industry after the 1880's that the national average of the iron content of ore varied little. "Little" is defined by the responsiveness of coke productivity to the iron content, estimated from a cross-section of British districts using rich and lean ores.[7] The curve "Corrected coke productivity" in Figure

6. Appendix D contains the details on the sources and methods for this.
7. Assuming that a higher ratio of pig iron made to ore used is an indication of greater ore richness alone (and not of a higher production

3 represents coke productivity as it would have moved had the iron content of ore stayed at its 1880 level throughout the period. Aside from a lower level after 1880, its pattern is identical to that of the uncorrected measure.

The other possible source of divergence between the average product of coke and a true productivity measure would be a decline in the ratio of the other inputs, capital and labor, to coke. Again, however, the evidence for stagnation after the mid-1880's in pig iron is left unaltered. If both capital and labor had become cheaper relative to coke from the 1880's down to the war, ironmasters would have substituted capital and labor for coke and this event, if anything, would have driven up the average product of coke. In this case the observed failure of the average product to rise would be definite evidence that the production function of pig iron had not risen over the period. History has been accommodating, for the prices of both capital and labor did fall relative to the price of coke after the 1880's, as might have been expected in view of the diminishing returns in British coal mining.[8] In other words, it is unlikely that substitutions of coke for other factors held down the average product of coke: the stagnation after the

---

function), a 1 percent rise in ore richness was associated with a rise of coke productivity among seven districts of about 0.7 percent in 1887, 1.0 percent in 1897, and 2.0 percent in 1912. The method is crude, but any coefficient in this range would yield similar results.

8. The coke price per ton and the yearly wage that are used below in the total productivity measure are reliable indicators of trends. The wage fell relative to the price of coke from the 1880's to 1920. The cost of capital is the price of capital goods multiplied by the sum of the rates of interest and depreciation. The price of capital implicit in the share of capital series used later to construct the total productivity measure is probably unreliable, so one must look elsewhere. The price index of capital goods implicit in C. H. Feinstein's real net investment series for the United Kingdom (Mitchell, *Abstract of British Historical Statistics,* p. 373) falls markedly relative to the price of coke from the decade 1885–94 to 1905–14. The slight rise in the yield on consols between the same decades is some offset. For I. L. Bell's firm in the Cleveland district the frequency of blast furnace relinings, a rough indicator of the rate of depreciation, exhibits no trend. In short, the price of capital fell relative to the price of coke from the 1880's to World War I.

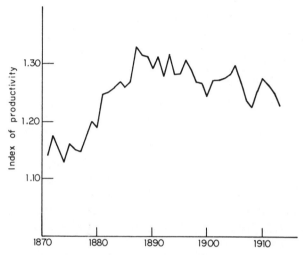

Fig. 4. Total Productivity in Pig Iron, 1871–1913

mid-1880's was a true stagnation of productivity in the pig iron industry.

As has been mentioned, the coke rate, corrected as it is here for the iron content of ore and the probable direction of substitution of labor and capital for coke, will prove useful for the international comparisons of productivity undertaken below. It is nonetheless a partial measure, and the full measure, comparing output with an index of all inputs of ore, labor, capital, and coke, is worth calculating as a check on its accuracy. Considering that coke was the major input of pig iron subject to technological change, it is no surprise that the full productivity measure, exhibited in Figure 4, has much the same pattern as the coke rate.[9] It peaks a year or two later in the 1880's, after a rise of about 1 percent per year for fifteen or twenty years, but thereafter stagnates as does the simpler measure. Given the margin of error in calculations of this sort, this is as much as can be affirmed with certainty.

9. See Appendix D.

Since the measures of productivity will have to be cruder for the other branches of the iron and steel industry, it is useful to explore briefly these margins of error in the full measure of productivity for this branch, in which the volume of information makes exploration fruitful. Any measure of productivity must make economic assumptions to have economic meaning and the assumptions must be reasonably correct. Two of the assumptions underlying the full measure, static profit maximization and free variability of capital, ensure that the industry was in long run equilibrium. There is little that can be said about the truth of these, except that the sharp, almost monthly, variation in the number of furnaces in blast suggests that the capital input was adjusted quickly to meet new conditions. A third substantive assumption, that the firms in the pig iron industry lacked market power, ensures that in equilibrium the factors of production were being paid their marginal product. Chapter 2 marshalled the evidence for the truth of this assumption in the market for output and, in any case, it can be shown that the maximum effect of market power on the productivity measure happens to be small.[10] The two remaining assumptions are formal.

10. In the notation of the previous footnote, the measure of the proportional productivity change, A^*, in some year is

$$A^* = Q^* - S_C C^* - S_O O^* - S_K K^* - S_L L^*.$$

The shares S_C, S_O, and so forth, appear in the formula because they are equal to the elasticities of output with respect to the particular input when the industry is in competitive equilibrium. The true measure, in other words, is

$$A_T^* = Q^* - \epsilon_C C^* - \epsilon_O O^* - \epsilon_K K^* - \epsilon_L L^*$$

where ϵ_C, ϵ_O, and so forth, are the elasticities. Market power, whether in the product or factor market, would tend to lower the share of the inputs other than capital (capital's share is calculated as a residual). The understatement of productivity growth is the true measure minus the actual measure:

$$A^*_T - A^* = (S_C - \epsilon_C) C^* + (S_O - \epsilon_O) O^* + (S_K - \epsilon_K) K^* + (S_L - \epsilon_L) L^*.$$

To find an observable upper bound to the understatement, make the extreme assumption that *all* capital's share comes from the profits of market power, but that the share minus the elasticity of other factors (such as $S_C - \epsilon_C$) is zero (rather than less than zero: allowing them to be less

82

It is assumed that the pig iron industry's production function exhibited constant returns to scale, that is, that the elasticities of output with respect to the inputs summed to one. Without external knowledge of the amount by which the sum exceeded one it is difficult at times to distinguish productivity change from returns to the scale of the industry as a whole.[11] It is hard to conceive of important channels of influence of scale, however, that are not fully reflected in the measured inputs. The final assumption, that the productivity change was neutral in its impact on each input, is still less critical: even when the shift in the production function is twisted, the productivity equation still measures the proportional change in output attributable to the shift.[12]

If the measure of total productivity is robust to violations of the economic assumptions involved, it is nonetheless true that estimating year-to-year productivity change in the pig iron industry involves compromises between the ideal and the available data, which introduce biases of their own. The chief difficulty of this sort is in the capital measure. In particular, the physical surface area of blast furnaces might be a better estimate of the capital per furnace than the estimate used here, the physical volume, and, if it were, the input of capital would grow slower and productivity

---

than zero would make the understatement still smaller). The understatement reduces, then, to $S_K K^*$, which is only about 0.06 from 1886–93 to 1912–20. In other words, the maximum correction for market power would be negligible.

11. G. T. Jones chose to call his remarkable book of productivity calculations in cotton, building, and pig iron (written in the 1920's but published later) *Increasing Returns: A Study of the Relation Between the Size and Efficiency of Industries, with Special Reference to the History of Selected British and American Industries, 1850–1910* (Cambridge,: Cambridge University Press, 1933). The identification of productivity change and increasing returns is a peculiarity of Marshallian economic vocabulary. Most modern studies of productivity change have found that when a distinction between the two is made the gain from increasing returns is by itself of slight importance.

12. See n. 5 above. The weighted sum of the biased rates of productivity change (λ^* and the like) will be equal to the measure assuming neutral productivity change.

faster. It happens, however, that the choice between the two estimates is unimportant. Much of the movement in the capital estimate as it stands is due not to the correction for the physical volume of blast furnaces but to the movement in the number of furnaces in blast. Moreover—and this applies to the labor estimates as well—large errors in the capital estimates result in small errors in the productivity index because the share of capital is small. The material-intensity of the industry and its relative capital-lightness make the difficulties surrounding the measurement of capital of negligible importance.[13]

There is solid evidence, then, that productivity change in pig iron was rapid before the mid-1880's and nil thereafter. Whether the cessation of productivity change was a failure of entrepreneurship or merely an exhaustion of available technique remains to be seen. At any rate the early growth (which was associated with the increase to the optimum in blast furnace heights) makes any allegation of British failure in the 1870's and early 1880's implausible. On the face of it, only in the 1890's and 1900's could entrepreneurs in British pig iron have fallen far behind their foreign competitors.

Bessemer Steel Rails. The evidence is less solid on the pattern of productivity change in steel. Although there have been no

13. After 1890 the volume of a typical blast furnace grew at a compound rate of growth of 1.81 percent per year, while its surface area grew at about 1.25 percent per year. The difference between these two, 0.56 percent per year, is the amount by which capital growth per year would be overestimated if a volume estimate were used when surface area was, in fact, the correct estimate. But an error in the growth rate of an input, like its true value, enters the formula for A^* weighted by its share in total costs. The shares of ore, coke, labor, and capital (capital is the residual share) averaged 0.45, 0.30, 0.10, and 0.15. Therefore, the underestimation of A^* due to an incorrect choice of the volume measure is the overestimate of the growth rate of capital weighted by capital's share: $(0.56) \cdot (0.15) = 0.084$. Correcting for this underestimate by compounding it over the fifty years would raise the index of the productivity level (the index of A) by only 4 percent by the last year: $(1.00084)50 = 1.042$. That is, it happens that using a surface area index of capital per blast furnace would raise the productivity measure only 4 percent in fifty years above its level using a volume index.

attempts in the literature to measure productivity in steelmaking, the emphasis on a heroic age of innovation during the late 1850's and early 1860's leaves the impression that later developments were negligible. The accomplishments of Bessemer, William Kelly, and Mushet, however, should not obscure the many later improvements in the Bessemer converter and its subsidiary processes, from a false bottom for the converter to an electric drive for the rolling mill. The difficulty with assessing the significance of these improvements is that direct measures of productivity in terms of the physical quantities of inputs and output are not feasible for Bessemer steel: the only significant and easily measured input, pig iron, was used, like ore in the making of pig iron itself, in a relatively fixed ratio to output. It was labor and capital that the innovations saved and the direct information on their quantities is poor.

The solution is to measure productivity with prices of inputs and outputs rather than with quantities. The quantity measure is merely a generalized average physical product, measuring the physical output of steel per unit of composite input. Looking at the matter this way suggests immediately that there might exist a similar measure that would be a generalized *marginal* physical product. There does: because the ratio of an input's price to the price of output is equal under conditions of competitive equilibrium to the input's marginal physical product, the price ratios in the new measure (the price measure) take the place of the quantity ratios in the quantity measure. Just as the quantity measure is calculated by subtracting a weighted average of the proportional changes in the quantities of inputs from the proportional change in the output, the price measure is calculated by subtracting the proportional change in the price of output from a weighted average of the changes in the prices of inputs. The quantity measure tells how much output would have increased if the quantities of inputs had not changed, the price measure how much price would have fallen if the prices of inputs had not changed. The price measure, indeed, is on the face of it a more directly useful concept in an analysis of supply and demand, for it measures directly the extent of movement in the supply curve. The difference in concept is, however, less significant than it appears, for under constant returns to scale (which must in any case be assumed to justify either measure) the price and quantity measures of productivity must be equal. This equality between the two measures is not surprising, considering that in the case of one input there are constant returns to scale if and only if the marginal and average products are equal.[14] The equality of the two, it should be noted, is exact, not approximate. The price measure is no more volatile or uncertain than the quantity measure: with consistent data it is identical to it.[15]

Since the major inputs into Bessemer steel rails were pig iron and labor, an estimate of productivity change in the making of rails, using the price measure, should include these. Thus,

$$A^* = S_I(P_I{}^* - P_R{}^*) + S_L(w^* - P_R{}^*),$$

14. This is a trivial special case of a duality theorem first proven by R. W. Shephard in *Cost and Production Functions* (London: Oxford University Press, 1953). The theorem can be used to make the following assertion. If $Q = A_q \cdot F(K, L)$ is a production function with a neutral technical change parameter, A_q, and with F exhibiting constant returns to scale, then there exists an average cost function, $P_q = A_p \cdot G(P_L, P_K)$, where G also exhibits constant returns to scale and where $A_q{}^* = -A_p{}^*$. That is, the rate of rise of the production function attributable to technical change is equal to the rate of fall of the cost function attributable to technical change.

15. The meaning of "consistent" is that the price and quantity data are consistent with the input shares used to weight them. D. W. Jorgenson pointed out that equality is a consequence of consistency in his "The Embodiment Hypothesis," *Journal of Political Economy*, 74 (1966), 3n. The argument is straightforward. Revenues are equal to costs, that is, using the standard notation, $PQ \equiv wL + rK$. Taking proportional rates of change of this identity,

$$P^* + Q^* \equiv \frac{wL}{wL + rK}(w^* + L^*) + \frac{rK}{wL + rK}(r^* + K^*).$$

whence

$$Q^* - S_L L^* - S_K K^* \equiv -P^* + S_L w^* + S_K r^*.$$

The left-hand side is the quantity measure and the right-hand side is the price measure.

where $P_I{}^*$ and w^* are the proportional changes in the prices of pig iron and labor, S_I and S_L are their shares in total cost, $P_R{}^*$ is the proportional change in the price of Bessemer steel rails, and A^* is the proportional change in productivity. Significant savings of pig iron, labor, or other inputs from the point where the pig iron arrives at the Bessemer converter to the point where the rail comes out of the finishing mill ready for sale will register in this measure.

The most important (as well as the most accessible) input price is the price of pig iron and Figure 5 exhibits the logarithm of its ratio to the price of rails. The slope of this graph is the propor-

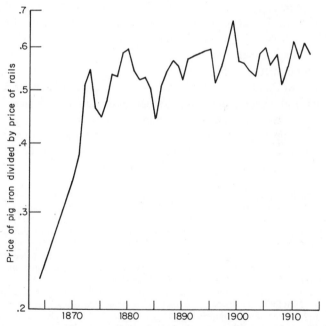

Fig. 5. The Marginal Product of Pig Iron in the Making of Bessemer Steel Rails, 1864–1913

87

tional change in the marginal physical product, that is to say, the term $(P_I{}^* - P_R{}^*)$ in the productivity measure.[16] Taking into account the wage term in the measure alters this pattern of productivity very little. Until the end of the 1880's the wage data for the iron and steel industry are poor, the only branch of the industry with good data being puddled iron. The ton-wages of the highly skilled puddlers, unfortunately, are unlikely to correlate well with the hourly wages of steelworkers, who were less skilled, particularly in view of the probable difference in the movement of the derived demands for labor in the contracting trade of puddled iron compared with the expanding trade of Bessemer steel. That is, scarcity rents for puddlers were probably dissipating as steel became competitive with iron. Before 1890, in fact, puddlers' wages fell, while wages in other industries rose: between the 1870's and the 1890's in agriculture, building, cotton textiles, coal mining, and engineering wages rose by at least 10 percent, while wages in puddled iron fell at least 10 percent.[17] Before 1890, therefore, steelworkers' wages are estimated here on the basis of shipbuilding and engineering wages rather than puddling wages. The data are to be weighted by the share of labor in costs, but it is difficult to find information on the share. Fortunately, the result can be shown to be insensitive to wide variations in the share. True productivity change between two years is bounded by the

16. The figure represents the West Coast price of low phosphorous pig iron divided by the West Coast price of Bessemer steel rails (Appendix B describes the sources for these prices). Low phosphorous pig iron is the relevant input because the acid process required it, and most Bessemer steel was made by the acid process. The rail price is not in fact given in the Appendix, since it runs parallel with the North Yorkshire price given there and has the same source, i.e., the weekly numbers of the *Iron and Coal Trades' Review,* through most of its length.

17. The wages other than in puddling are the Bowley and Wood indexes given in Mitchell, *Abstract of British Historical Statistics,* pp. 348–351. The shipbuilding and engineering wage index, for example, rises from about 82 in 1860–1869 to about 101 in 1890–1899. Wages in puddling are given in *Returns of Rates of Wages,* pt. II, "Iron and Steel Manufacturers," S.P., 1887, vol. 89, pp. 146–156, and in S.P., 1900, vol. 82, p. 572.

proportional change in the marginal products of pig iron and of labor: the one is the productivity measure if all costs were pig iron and the other if all costs were labor. If one ton of pig iron produced as much as one ton of steel (since no scrap was used in the Bessemer process this is an upper bound), the share of pig iron would be, at a minimum, simply the price of one ton of pig iron divided by the price of one ton of steel, P_I/P_R. At a maximum, then, the share of labor would be the proportion of costs remaining from this minimum share of pig iron, $1 - (P_I/P_R)$. The price of pig iron and the wages of labor move at so much the same rate, it happens, that any mixture of these crude extremes yields much the same result: the pattern of the productivity measure in Bessemer railmaking is little affected by the most perverse variation in the shares, and its outline is that of the marginal product of pig iron alone.

The two notable features of productivity in Bessemer steel rails uncovered by the measure are its rapid growth in the 1860's and 1870's—over 5 percent a year—and its stagnation from the early 1880's on. The significance of this stagnation, like the stagnation in pig iron, will not be clear until productivity in Britain is compared in the next chapters with productivity abroad, but the early growth is directly relevant to an understanding of the adoption of steel. For twenty-five years after Bessemer first announced his steelmaking process to the world it continued to mature; as will be shown in a moment, a similarly long period of maturation characterizes the Siemens open hearth process. Many observers, however, concentrating on the earliest history of each steelmaking process, assume that the pattern of Athena springing full-grown from the brow of Zeus, rather than the one of gradual maturation, fits the major innovations in steelmaking. Consequently, they overlook growing productivity as an explanation of the falling price and growing output of the new steels. Thus, J. H. Clapham, listing the "limitations to a very rapid extension of the mass uses" of Bessemer steel, implicitly assumes that the productivity of the process did not increase in the 1860's and 1870's. The limitations, he says, were: "First, [Bessemer's] high charge for patentee's licences . . . Second, the relatively expensive 'Bessemer' raw material . . . Third, the year-long tests of endurance on the permanent way, in the boiler or the ship, necessary to prove that the dearer material was really the more economical. [Fourth], the gigantic lock-up of capital and human skill in puddling, and [fifth] the undisputed dominance of the British iron industry in the world's markets."[18] The price of steel rails, however, fell from about 350 shillings per ton in 1864 to 118 shillings per ton in 1880, while the price of pig iron to be made into steel actually rose slightly. Clapham's argument carries little conviction in the face of such an enormous fall induced by productivity change in the supply curve of Bessemer steel. The royalty on the process, for example, amounted to about 25 shillings a ton before its reduction in 1870 to 2.5 shillings,[19] hardly enough to explain a fall in price of 240 shillings. Indeed, unless the supply of steel were perfectly elastic or demand perfectly inelastic, by no means all of the 25-shilling reduction in cost would be transmitted to the price.[20] The main limitation to a very rapid extension of the mass use of Bessemer steel was the crudeness and expense of the process in its early form. As the process matured, its price fell dramatically and its use increased.

Open Hearth Steel Ship Plates. The open hearth process experienced a similar period of maturation, extending from Siemens's earliest successful experiments in the late 1860's to the adoption of tilting furnaces in the early 1900's. Its history can be sketched

18. *Economic History of Modern Britain,* vol. II, p. 56.
19. In January 1870, the *Iron and Coal Trades' Review* reported that the effective royalty was then 22.5 to 27.5 shillings a ton and spoke of its expected fall to 2.5 shillings when Bessemer's main patent expired in February (19 February, 1870, p. 40). There is confirming evidence that this was the size of the royalty (see Burn, *Economic History of Steelmaking,* p. 22).
20. If the elasticities of supply and demand were numerically equal, for instance, only half of the reduction in royalty would have been reflected in the price.

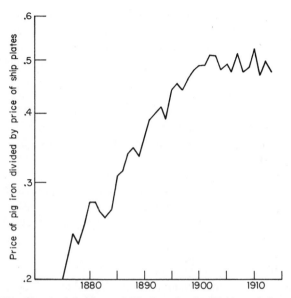

Fig. 6. The Marginal Product of Pig Iron in the Making of Open Hearth Ship Plates, 1872–1913

quickly, for the logical underpinnings of the productivity measures involved have already been adequately exposed to view in the treatment of the other branches of iron and steel. The marginal product of pig iron in open hearth ship plates, shown in Figure 6, grew at about 3.0 percent per year from the late 1870's to the early 1900's and that of labor at 4.2 percent per year.[21] This, like the rates of over 5 percent per year experienced in Bessemer steel before 1880, is very rapid productivity growth by compari-

21. The wage used is the Bowley-Wood index of weekly earnings in shipbuilding and engineering, between 1877 and 1902 (given in Mitchell, *Abstract of British Historical Statistics,* p. 350). The index generally correlates well with the scattered information on steel. Compare Sidney Pollard, "Wages and Earnings in the Sheffield Trades, 1851–1914," *Yorkshire Bulletin of Economic and Social Research,* 6 (1954), 48–64, esp. pp. 62–63. The sources for the prices of steel and pig iron are Appendix B.

91

son with the rates typical of the economy as a whole. It was shown earlier that the British economy experienced productivity change of a little over 1 percent per year in the late nineteenth century. A figure comparable with those in Bessemer railmaking before 1880 and in open hearth platemaking before 1902, moreover, would be still smaller: each industry's productivity change (which includes savings of inputs from other industries) is weighted by the ratio of its gross output to national income and is cumulated to yield the national rate of productivity change; since gross output is greater than national income for any given degree of industrial detail, the weights sum to more than one; therefore, the prevailing rate of productivity change for individual industries, taking into account their savings of inputs from other industries, must have been lower than the national rate by a factor equal to the ratio of national income to gross output. The inference is that productivity in open hearth platemaking was growing down to the early 1900's at a rate at least on the order of four times that prevalent elsewhere in the economy. Rolled products of open hearth steel accounted in 1907 for roughly half of the iron and steel industry's output of finished products. In consequence, even though productivity change was probably nil throughout the late nineteenth century in the older processes of the industry, such as cast and wrought iron, and ceased in Bessemer steel after the early 1880's and in pig iron after the late 1880's, productivity change in the industry was probably higher than elsewhere in the economy down to the first few years of the twentieth century.

It has sometimes been supposed that entrepreneurs in British iron and steel began to perform poorly in the 1870's. The record of productivity change in the industry, however, was good by national standards until the early 1900's. If there was a failure in the Victorian iron and steel industry, it was on the face of it late Victorian, not to say Edwardian, in timing. Even if the sustained record of competence in open hearth steelmaking is ignored, there is no case for failure in the rest of the industry until the late 1880's, and the supposed failure would not have generated a significant gap between the level of foreign and British productivity until the 1890's. There is little warrant, in short, for fixing the date of a failure of entrepreneurs in the early 1880's and none at all for fixing it in the 1870's.

6   Was Productivity Change More Rapid in
the American Industry?

To put the findings of the last chapter in a way that is less favorable to British entrepreneurs, by the first decade of the twentieth century productivity growth had ceased in the iron and steel industry. One might infer from this that at least after 1900, and perhaps before, entrepreneurial vigor flagged and Britain fell behind in productivity. It would be hazardous, however, to draw such an inference. The difference in the timing of the climacteric in different parts of the British industry casts doubt on it at once, for if failing vigor affected all entrepreneurs in the industry at the same time, as one might expect it would, it is puzzling that the climacteric in steel plates occurred twenty years after that in steel rails. Looking outside Britain, this chapter develops a still more compelling objection to the hypothesis of entrepreneurial failure, namely, that in much of the industry during much of the period 1870–1913 productivity in Britain grew as fast as it did abroad. If it did, one cannot argue that British entrepreneurs exhibited unusual ineptitude.

Productivity Change in American Iron and Steel. The task is to separate the growth in the world's technological knowledge from variations in the performance of British entrepreneurs, and for this purpose the best point of comparison is the American industry. The French, the Belgians, or more doubtfully, the Ger-

94

mans might have deviated from time to time from best practice, but it is usually supposed that the Americans, exhibiting "engineering common sense which amounts almost to genius,"[1] never did: American practice was the international standard of perfection in iron and steel as in many other industries from the 1880's on. The statistical record of the American industry, moreover, is very full, the major sources being the weekly *Iron Age* for market prices, the yearly statistical volumes of the American Iron and Steel Institute for quantities of output, and, for more intimate details on inputs and costs, the decennial (quinquennial from 1900) census of manufacturers.

The censuses for 1889 and 1914 (1889 is the earliest year for which the census data are fully comparable with later censuses) bracket the years during which American practice was most likely to have surpassed British practice. The information in no census is complete, and to bring to bear relevant outside information, which consists primarily of prices, it is convenient to use the price measure of productivity change developed earlier, that is, to subtract the proportional change in the price of output from the proportional change in a weighted average of the prices of inputs.

The one difficulty not already encountered earlier in handling the less rich British data is the inhomogeneity of the industry's output. The American census separates the industry into two major branches, Blast Furnaces and Steelworks and Rolling Mills. The various outputs of blast furnaces (that is to say, the types of pig iron) were highly substitutable for one another in both production and consumption, and present therefore no problems of aggregation. The output of steelworks and rolling mills, on the other hand, was less homogeneous and requires an aggregation procedure. The natural one is to weight the proportional changes in the prices of rails, bars, plates, sheets, and other products by their shares in the value of output, because these weights, as was

1. As J. W. Hall, an English steelmaker, said in 1902. He is quoted in Burn, *Economic History of Steelmaking,* p. 184.

argued implicitly in Chapter 1, reflect, as it were, the national income embodied in each product.[2]

An index of output prices constructed in this fashion in steelworks and rolling mills (as distinct from blast furnaces), using nine categories of output, fell from 1889 to 1914 by 40.6 percent.[3] The price of inputs fell, too, but by much less than the price of outputs, indicating that there was indeed productivity change in American steelmaking. For inputs accounting for about 79 percent of costs the censuses give enough information to extract prices in 1889 and 1914, but for the remaining 21 percent of costs, including capital, natural gas, copper ingots, supplies, land rents, and taxes, outside estimates must be used. The calculation for the census inputs, which is similar to that for the price of output, is exhibited in Table 4. As the sum of the last column of the table shows, the weighted index of the input prices available from the censuses fell 9.3 percent from 1889 to 1914, so that the fall in the price of outputs unexplained by the fall in the

2. The procedure has a formal rationale. A "Divisia index" of output is an index of proportional changes in the various components of output weighted by their shares in the value of output. The Divisia index of output minus the Divisia index of input is a measure of the shift of the implicit transformation function of inputs and outputs, $F(Y, X) = 0$, if the usual assumptions of homogeneity, equilibrium, and competition are met. See D. W. Jorgenson, "The Embodiment Hypothesis" and D. W. Jorgenson and Zvi Griliches, "The Explanation of Productivity Change," *Review of Economic Studies,* 34 (1967), 249–283.

3. The categories were rails, bars, rods except wire rods, hoopbands, skelp, structural shapes, sheets, plates, other products of metal, and nonmetal products (the last two items made up 43 percent of the total by value, averaging the 1889 and 1914 shares in value). The price of an output was taken to be its total value divided by its quantity in tons. For certain items, making up an average of 6.5 percent of total revenue, there were no figures on the quantity of output, only values (by-products like fertilizer and gas fall into this category). Their price was assumed to have fallen as did the weighted average of the prices of the other products. Source: U.S. Department of Commerce, Bureau of the Census, *Census of Manufactures 1890, Part 3,* pp. 474–477 and *Census of Manufactures 1914. Volume II,* pp. 226 and 244 (Washington: Government Printing Office).

TABLE 4. Prices, Cost Shares, and Weighted Changes in the Prices of Inputs Making Up 79 Percent of Costs in American Steelmaking, 1889 and 1914.

(prices in dollars per long ton or per man year)

Input	Prices 1889	Prices 1914	Average of 1889 and 1914 shares in cost	Proportional change, 1889–1914, in price weighted by share[a]
Clerks	1490	1470	0.0276	−0.0003
Wage earners	540	760	.2140	+.0710
Iron ore	6.46	4.25	.0072	−.0030
Spiegeleisen	33.88	54.70	.0204	+.0096
Pig iron	17.36	13.55	.2540	−.0674
Scrap[b]	20.89	11.71	.0870	−.0488
Semifinished steel[b]	32.89	20.76	.1320	−.0599
Anthracite	1.74	2.46	.0030	+.0010
Soft coal[c]	2.09	2.40	.0382	+.0052
Coke[c]	3.74	3.48	.0030	−.0002
Sum			.7900	−.0930

Source: See n. 3, this chapter.

[a] The average of each item's share in 1889 and 1914 costs was multiplied by the proportional change in each item's price (namely, the absolute price change divided by the average of the 1889 and 1914 prices). Details may not add owing to rounding.

[b] Scrap and steel semifinished products are in this accounting *outside* (that is, either purchased on the open market or brought from another works of the same company). For 1914, at any rate, there is a similar accounting on the output side (for 1889 it is obscure).

[c] For 1914 the values of the fuel inputs were not given. They were extrapolated from the proportional relation prevailing between anthracite, soft coal, and coke used as blast furnace fuel and used as steel fuel in 1889. The Census of 1914, p. 225, gives the tonnages of fuels.

price of inputs depends on the price changes assumed for the unknown inputs in the following way:

$$A^* = 40.6 - 9.3 + (.109)P_k^* + (.104)P_0^*.$$

$P_k{}^*$ here is the proportional change in the cost of capital, 0.109 being its share in costs (observed as a residual from the other cost shares), and $P_0{}^*$ the proportional change in the price of miscellaneous items in the unknown category, 0.104 being its share. The measure of productivity, then, depends on the prices assigned to these two elements of cost, although not critically. With little significant error one might assume that the price of the miscellaneous items rose as wholesale prices did, 20.2 percent between 1889 and 1914.[4] The exchange value of capital goods fell about 2.8 percent and the rate of return on capital rose about 9.2 percent, resulting in a rise in the price of capital services $(P_k{}^*)$ of about 6.4 percent.[5] The rate of productivity change in steelworks and rolling mills implied by these assumptions about $P_k{}^*$ and $P_0{}^*$ is 34.1 percent from 1889 to 1914, or about 1.38 percent per year. It is hazardous, to be sure, to rely on two years alone to estimate productivity change in steel, but the data from the censuses can be checked against market data, which is available yearly. A productivity measure using factor shares from the censuses and market prices of steelmaking labor capital, pig iron, scrap, and miscellaneous materials for each year from the 1880's

4. The Bureau of Labor Statistics wholesale price index (with a small extrapolation from 1890 back to 1889 on the basis of the Warren and Pearson index) rises from 55.6 in 1889 to 68.1 in 1914. (*Historical Statistics*, p. 116).

5. The "exchange value of capital goods" is $P_k{}^e$ in the expression $P_k{}^e (r + d)$. This expression is the cost of capital services under certain simplifying assumptions. The proportional change in the cost of capital would be, then, $(P_k{}^e)^* + (r + d)^*$ (strictly speaking the cost of capital should include a term reflecting capital gains). The estimate of P_k is the price index used for deflating the book value of total capital (both fixed and variable) in American iron and steel and their products developed by Daniel Creamer, S. P. Dobrovolsky, and Israel Borenstein in their *Capital in Manufacturing and Mining* (Princeton, N.J.: National Bureau of Economic Research, 1960), namely, 56.5 in 1889 and 54.9 in 1914. The estimate of r, the opportunity cost of capital, is the yield on industrial common stock (Cowles Commission estimates, *Historical Statistics*, p. 656), namely, 0.044 in 1889 and 0.053 in 1914. The rate of depreciation is assumed to be 0.05 and unchanging.

to World War I yields roughly the same rate of productivity change, namely, between 1.1 and 1.5 percent per year.[6]

This is fairly rapid growth, although in the British steel industry as long as the improvements in open hearth practice continued (that is, down to the early 1900's) growth rates were in a similar range. As in the British industry, moreover, the experience of the Bessemer and open hearth branches of the American industry in the 1890's and 1900's differed radically. In particular, British performance appears to have been no worse than American in Bessemer steelmaking. Market prices in America and Belgium are complete enough to form estimates back to the 1880's of the marginal product of pig iron in steel railmaking (which was dominated by the Bessemer process), and these marginal products stagnated after the 1880's in America and Belgium just as they did in Britain.[7] The impression of worldwide stagnation is reinforced by data from the American censuses on the marginal products of pig iron, scrap, and labor in Bessemer production (the first three rows of Table 5). The rise in the American marginal product of labor in Bessemer rails (the second row of the table) was

6. The range is generous. The rate based on market data (metal inputs and outputs from *The Iron Age* and other prices from various sources, for example P. H. Douglas's series of annual earnings in manufacturing, given in *Historical Statistics,* p. 91) is about 1.23 percent per year.

7. The American prices for this comparison are those given in Temin, *Iron and Steel in America,* Appendix C, Tables C.15 and C.15A, pp. 283–285. The Bessemer steel rail price is a price at the works in Pennsylvania from 1867 to 1911. To compute the pig iron to rail price ratio before 1885 I have used the price of #1 foundry pig iron at Philadelphia. This price gives results similar to the Bessemer pig iron price after 1886. Before 1886 one would expect the price ratio of Bessemer pig iron to foundry pig iron to have fallen, if it changed at all, because the Bessemer-quality lake ores would have become cheaper as transport improved. In this case using the foundry pig iron price would tend to exaggerate the rate of growth of productivity in the American rail trade. Correcting for the exaggeration would reinforce, not weaken, the conclusion that the American rate of productivity growth was no greater than the British rate. The Belgian data are from C. Reuss, E. Koutny, and L. Tychon, *Le Progrès Economique en Sidérurgie: Belgique, Luxembourg, Pays-Bas, 1830–1955* (Louvain and Paris: 1960), pp. 397–398 and pp. 405–406.

TABLE 5. The Marginal Products of Inputs of Pig Iron, Scrap, and Labor in American Steelmaking, Census Years, 1879–1914.
(price of input divided by price of output, yielding units of tons of output per ton of material input or per man year of labor)

Products	1879	1889	1899	1904	1909	1914
Bessemer products						
Pig iron/rails	0.46	0.56	—	0.53	0.57	0.53
Labor/rails	7.70	17.00	27.00	22.00	25.00	28.00
Pig iron/rods	0.35	0.42	0.36	0.41	0.49	0.43
Open hearth products						
Pig iron/structural shapes	.21	.34	.46	.41	.50	.49
Scrap/structural shapes	—	—	.46	.38	.49	.42
Labor/structural shapes	3.50	9.30	16.00	17.00	22.00	27.00
Pig iron/boiler plates	0.31	0.30	0.31	0.34	0.39	0.41
Scrap/boiler plates	—	—	.35	0.31	.38	.33
Labor/boiler plates	5.10	8.80	12.00	14.00	17.00	24.00
Pig iron/black plates	—	—	0.27	0.28	0.32	0.34
Scrap/black plates	—	—	.30	.26	.31	.27

Sources: 1879: *Tenth Census of the U.S. Final Report,* vol. II, *Report on Manufactures* (Wash., D.C.: U.S. Government Printing Office, 1883); 1889: *Eleventh Census,* vol. VI, *Report on Manufacturing,* pts. 1 and 3 (Wash., D.C.: 1895); 1899: *Twelfth Census,* vol. IX, *Manufactures, Special Reports on Selected Industries* (Wash., D.C.: 1903); 1904: *Census of Manufacturers, 1905,* pts. 3 and 4, *Special Reports on Selected Industries* (Wash., D.C.: 1907); 1909: *Thirteenth Census,* vol. X, *Manufactures, 1909, Reports for Principal Industries* (Wash., D.C.: 1913); 1914: *Census of Manufactures, 1914,* vol. II, *Reports for Selected Industries* (Wash., D.C.: 1918).

partly matched in Britain, and in any case labor was a relatively unimportant input into rails. Pig iron was the major input, and the small rise in its marginal product (the first and third rows) indicates that productivity change in the Bessemer branches of the American steel industry was slight after the 1880's, so slight that there could develop no large divergence indicative of entrepreneurial failure between American and British performance in Bessemer steel.

American superiority in steelmaking, therefore, could develop only in the open hearth branch. The marginal products of pig iron and labor exhibited in the table confirm the implication of the earlier measures that down to 1899 American performance was not superior: the average of the rates of change of the marginal product of pig iron in structural shapes and boiler plates was roughly 2.0 percent per year from 1879 to 1899, lower than the comparable British rate, and the average of the marginal product of labor 6.0 percent per year, higher than the British rate; in view of the larger share of pig iron than labor in total costs, these imply if anything a faster rate of change in Britain than in America. After 1899 the marginal products of inputs to the open hearth in America, however, continued to grow, although not at very fast rates. The marginal product of labor did rise substantially down to 1914, but in structural shapes the marginal product of pig iron did not, and in all open hearth products the marginal product of scrap, which was a more important input in America than in Britain, stagnates. A divergence between American and British levels of productivity, in short, could have developed only in the decade or so before World War I, and even then the earlier specialization of the British industry on the open hearth and the rapid rates of productivity change accompanying it suggest, as the next chapter will confirm, that continued productivity growth in America after 1899 was the result of a British lead before 1899.

The period of time during which American productivity growth was more rapid than British was considerably longer in pig iron than in steel. The price of pig iron reported in the censuses fell 18.5 percent from 1889 to 1914, while the prices of inputs, shown in Table 6, rose slightly. The implied rate of productivity change in American pig iron is 0.75 percent per year from 1889 to 1914,

TABLE 6. Prices, Cost Shares, and Weighted Changes in the Prices
of Inputs in American Pig Iron, 1889 and 1914.
(prices in dollars per ton or per man year)

Input	Prices 1889	1914	Average of 1889 and 1914 shares in cost	Proportional change, 1889–1914, in price weighted by shares
Clerks and officers	1510	1590	0.0150	+0.0008
Wage earners	437	776	.0860	+.0480
Domestic iron ore	4.10	3.40	.4200	−.0784
Foreign iron ore	6.07	5.41	.0350	−.0040
Mill cinder, etc.	2.70	3.07	.0205	+.0027
Fluxes (limestone, etc.)	0.84	0.97	.0320	+.0049
Coke	3.33	3.48	.2260	+.0102
Soft coal	1.54	1.77	.0022	+.0004
Anthracite	2.88	4.06	.0180	+.0073
Charcoal ($/bu.)	0.75	0.58	.0180	−.0047
Miscellaneous inputs (index of price)	55.60	68.10	.0480	+.0097
Capital (index of price)	53.00	56.50	.0775	+.0050
Sum			1.000	+.0019

Source: Census of Manufactures, for the indicated years, as in n. 3, this chapter. The prices for the two last categories are assumed to be the same as in steel, as in nn. 4 and 5, this chapter. The share of capital was taken as a residual from the other inputs.

which is lower than in steel but high enough in view of the stagnation of productivity in British pig iron after the late 1880's to produce eventually a substantial divergence between British and American performance. By 1914 American productivity would have been about 20 percent higher than British productivity, had the two been equal in 1889.

It might be noted here that G. T. Jones in the 1920's estimated American productivity growth in pig iron from 1889 to 1914 at a higher rate, about 1.26 percent per year, which would have produced a higher divergence of 37 percent by 1914.[8] Jones's work was extraordinarily precocious in conception, for his measure of "increasing returns" is exactly the price measure of productivity change.[9] The high rate of productivity change in the American

8. G. T. Jones, *Increasing Returns*, p. 296, col. 3. His results have been widely and somewhat uncritically accepted; for example, by Clapham, *Economic History of Modern Britain*, vol. III, p. 70.

9. Jones's index of what he calls "real costs" (denoted by p_a')—by which he means the cost of pig iron that would have been observed had the prices of inputs not altered from their levels in the base year—is defined in his notation (*Increasing Returns*, p. 33), as

$$p_a' \equiv p - \sum_i [\frac{Y_{ai} (\pi_i - A_i)}{X_a}].$$

p is the price of pig iron actually observed in the given year for which real costs are to be calculated. Y_{ai} is the amount of input i used in the base year a, X_a the amount of pig iron produced in the base year, π_i the price of the input i in the given year and A_i its price in the base year. It will be easier to understand this definition of real costs if it is translated into the notation that has become standard since Jones wrote and if it is made concrete by using only two inputs (capital, K, and labor, L, with their prices r and w) producing output, Q, with a price P. If P_t' is the real cost in year t and year 0 is the base year, the translation yields:

$$P_t' \equiv P_t - [\frac{L_0(w_t - w_0)}{Q_0} + \frac{K_0(r_t - r_0)}{Q_0}]$$

A hard look at this expression will convince the reader that it does indeed, as Jones asserted, define the price that would have been observed had input prices not altered from the base year: the labor and capital requirements per unit of output in the base year are multiplied by the observed changes in the prices of labor and capital, and the result—which is the contribution

industry that he found, however, is explained by certain flaws in the execution of the measure. He used, for example, coke prices at the ovens and ore prices at the mines rather than at the blast furnace: because the price of transport and handling was falling rapidly, the prices used understated the fall of input prices and, consequently, led to a large overstatement of productivity change.[10] The divergence of 20 percent by 1914 is the correct one.

The potential divergence between British and American productivity in pig iron as in open hearth steel, however, depends critically on the initial relative levels of the two, which the next chapter will measure. At any rate, the question in the title of the present chapter—was productivity change more rapid in the American industry?—must be answered period by period. In Bessemer steel the answer is a clear "no," but in open hearth steel the American industry had more rapid productivity growth for the last ten or so years of the forty-year period and in pig iron for the last twenty.

---

of changes in the prices of inputs alone to the change in the price of output —is subtracted from the observed price. That this expression is identical to the price measure of productivity is now easily proven. Subtract P_0 (the price of pig iron in the base year) from both sides of the identity, divide both sides by P_0, and multiply the term on the right-hand side associated with labor and capital by w_0/w_0 and r_0/r_0. Collecting terms yields:

$$\frac{P_t' - P_0}{P_0} \equiv \frac{P_t - P_0}{P_0} - [\frac{w_0 L_0}{P_0 Q_0} (\frac{w_t - w_0}{w_0}) + \frac{r_0 K_0}{P_0 Q_0} (\frac{r_t - r_0}{r_0})]$$

or, in the notation used earlier,

$$P_t'^* \equiv P_t^* - s_L w_t^* - s_K r_t^*.$$

That is to say, the rate of change of Jones's index of real costs between year o and year t (i.e., $P_t'^*$) is identically equal to the negative of the price measure of productivity change (since $A^* \equiv -P_t^* + s_L w_t^* + s_K r_t^*$).

10. The Census costs of coke used in blast furnaces, for example, reflect the transport and handling costs between the coke oven and the gate of the pig iron works. In 1889 the Census value exceeded Jones's coke-oven price by \$1.80; \$1.80 agrees with the costs of assembly given by C. D. Wright (Wright, however, was also in charge of the Census) in his study of American blast furnaces in 1888–89 (*6th Annual Report of the Commissioner of Labor, Costs of Production: Pig Iron, Steel, Coal, etc.* [Washington, D.C.: 1891]).

Slow Growth and Antique Technology in Steel. Although the observed rate of foreign productivity growth does not provide a conclusive measure of the alleged divergence between British and foreign performance, it does provide a strong argument against one major explanation of it in the steel branch of the industry. The explanation, reaffirmed by successive generations of observers of the industry, hinges on the slow growth of demand in British markets compared with the fast growth of the protected markets of America and Germany, slow growth which left Britain with old capital equipment embodying an antique technology.[11] Its popularity among contemporaries and historians is attributable perhaps to its uncluttered logical structure and its air of mild paradox: as Peter Temin put it recently, "high costs in Britain appear . . . as a result of slow growth, not its cause."[12] Until Temin's work, the argument had little besides its aesthetic qualities to recommend it. His contribution was to cast it into quantifiable and refutable form, building on the recent economic literature on embodied technical change.[13] In applying the argument to the steel industry, he found that the divergence in productivity between the slowly growing British industry and the rapidly growing American and German industries must have been, by 1913, on the order of 15 percent. If the demand effect on productivity and prices was as large as 15 percent it would certainly warrant the emphasis it has received in the historical literature, and would undermine the argument being developed here that a divergence between British and foreign levels of productivity did not in fact

11. See the works cited in Chap. 1 above, n. 22. Compare Burn, *Economic History of Steelmaking*, p. 185: "The rate of modernization was as certainly less in Great Britain than in the states or on the Continent as the volume of innovation was less . . . This in some degree would be a virtually inevitable result of the growing home markets of foreign rivals."

12. "The Relative Decline of the British Steel Industry, 1880–1913" in H. Rosovsky, ed., *Industrialization in Two Systems*, p. 151.

13. Robert Solow's "Investment and Technical Progress," pp. 89–104 in Kenneth Arrow et al., *Mathematical Methods in the Social Sciences 1959* (Stanford: Stanford University Press, 1960), is the beginning of a large literature on the "embodiment" argument.

exist. Temin's estimate, however, which is rapidly finding its way into other works on the late Victorian economy,[14] contains serious upward biases. Correcting the biases yields an estimate of a productivity difference in 1913 not of 15 percent but less than 1 percent.

Temin gives a clear outline of the rationale of the calculation (he compares Britain with Germany, but a comparison with America is implied as well and would give very similar results):

> Under certain equilibrium conditions, the average age of capital in an industry is equal to the reciprocal of the sum of the rate of growth of the industry and the rate of depreciation in force, both being constant exponential rates . . . If the rate of depreciation was 5 percent a year in both countries [and if the rate of growth of the industry was 3.4 percent in Great Britain and 9.6 percent in Germany], the formula indicates an average age of twelve years for the capital in the British steel industry and seven years for the capital in the German industry. If the rate of embodied technological change were about 3 percent a year, a five-year difference in the average age of capital would mean a difference in costs of about 15 percent.[15]

The first upward bias arises from the implicit assumption that the appropriate difference in the ages of German and British capital is the long-run equilibrium difference, 5 years by Temin's calculations. As he recognizes, however, the traverse from one equilibrium age to another takes time: "The difference between the average age of capital in the two countries to be expected from their differing growth rates then grew from near zero in 1890 to about five years in 1913."[16] It can be shown that at any time T years after the change in the growth rate of output the average age (a_T) will depend on T, on the old and new rate of growth (q^* and q), and on the old and new rate of depreciation (δ^* and δ) in the following manner:

14. Such as S. B. Saul's *The Myth of the Great Depression* (London: Macmillan, 1969), p. 49; and Payne, "Iron and Steel Manufactures" in Aldcroft, ed., *The Development of British Industry,* p. 97.
15. Temin, "Relative Decline," p. 150. The phrase in brackets is implied.
16. Ibid., p. 150n.

$$a_T = (\frac{1}{q^* + \delta^*}) (\frac{e^{-\delta T}}{e^{qT}}) + (\frac{1}{q + \delta}) (\frac{e^{qT} - e^{-\delta T}}{e^{qT}}).$$

A more transparent expression of this fact is

$$a_T = a^* (\frac{D}{Q}) + a (\frac{Q - D}{Q});$$

that is, the age of capital T years after the change is a weighted average of the old equilibrium age a^* and the new equilibrium age $a,$ where the weights depend on the exponential functions of time, D and Q.[17] Temin asked what the difference in age between

17. Let τ^* be the year (namely, 1890) in which the age of capital in the British industry started to rise as a consequence of the new lower rate of growth of output (denoted by q, distinguished from the pre-1890 rate, q^*). Let a^* be the average age of capital in 1890 (equal to $\frac{1}{q^* + \delta^*}$, presuming that the age was in equilibrium) and let K_{τ^*} be the amount of capital in existence in 1890. The age of the surviving parts of this capital in some year t after 1890 (τ^*) will be, obviously, $a^* + t - \tau^*$: if a^* was, say, 3 years in 1890, with each passing year it would age a year, its total age being 13 years in, say, 1900 ($= t$). Notice that $t - \tau^*$ is the T given in the formula in the text. The old capital surviving in the year t will be $K_{\tau^*} e^{-\delta (t - \tau^*)}$: it depreciates continuously at the rate δ. The new and up-to-date capital created in some year t' after 1890 will be $(q + d)$ $(K_{\tau^*} e^{q (t' - \tau^*)})$, that is to say, the rate of gross investment (q being the rate of net additions to capacity and δ the rate of replacements) multiplied by the stock of capital in existence in year t'. From the year t' down to the year t (for which the age of capital is to be measured) this new capital ages $t - t'$ years and depreciates at the rate δ. The aggregate years of age in the capital stock put in place after 1890 (that is, the amount having each age multiplied by that age) and surviving to year t will be the sum of the aggregate years of age for all vintages, appropriately depreciated, from τ^* to t:

$$\int_{\tau^*}^{t} [t - t'] [(q + \delta) K_{\tau^*} e^{q(t - \tau^*)}] [e^{-\delta (t - t')}] dt'.$$

The first bracketed term in the integral expression is the age of capital of vintage t' in year t, the second the amount of that capital initially created, and the third the proportion of it surviving to year t. The comparable expression for pre-1890 capital is $(a^* + t - \tau^*) K_{\tau^*} e^{\delta (t - \tau^*)}$. Therefore, the average age in t is simply the sum of the aggregate age of post- and pre-1890 capital (the two expressions just mentioned) divided by the

the German and British steel industries would have been in 1913 given, first, that the growth rate of British output fell sharply after 1890, second, that the British capital stock had reached its new long-run average age (11.9 years on his assumptions) by 1913, and third, that the German growth rate did not recede from its high level at any time from 1880 to 1913. Under these assumptions, the difference in age would have been 4.85 years.[18] The last two assumptions, however, are incorrect. According to the formula above, the average age of British capital would have risen to 11.2 years by 1913 rather than to its ultimate equilibrium of 11.9 years. Moreover, as Temin remarks, "if the growth rate has changed in the recent past, the current growth rate is the most important rate for the determination of the average age of capital."[19] The growth rate of German steel fell sharply after 1900, and by 1913 the average age of capital would have been 7.55 years (according to the formula) rather than 6.77 years on the mistaken assumption of no change in the rate of growth from 1880 to 1913.[20] The age difference in 1913 is reduced to 3.6 years. So different was the German experience before and after 1900 that the age difference in 1900 would have been nearly as

amount of capital in existence in year t (which is $K\tau^* e^{q(t - \tau^*)}$). After some manipulations, which need not be reproduced here, the formula for the average age in year t reduces to that given in the text. Letting $T (= t - \tau^*)$ approach infinity shows that the ultimate equilibrium age is indeed $\frac{1}{q + \delta}$. R. R. Nelson derives a similar formula for the average age at t using discrete rather than continuous mathematics ("Aggregate Production Functions and Medium-Range Growth Projections," *American Economic Review,* 54 [1964], 585n), but the derivation here is simpler.

18. This is below Temin's estimate of 5 years because it replaces an erroneous growth rate of 9.6 percent per year between 1890 and 1913 in the German industry with the correct 9.17 percent per year. Throughout the discussion I use Temin's steel ingot figures (Temin, "Relative Decline," p. 143) and his assumption that output grew steadily between the years 1880, 1890, 1900, 1910, and 1913.

19. Ibid., p. 150n.

20. The comparison between Britain and America, as was asserted earlier, gives similar results. The growth rate fell off after 1900 in the United States as it did in Germany. The average age of capital in 1913 would have been a little under 7.4 years.

great as in 1913: the age of British capital would have been about 9.7 years by 1900, instead of its final equilibrium level of 12.9 years, but the German age would have been 6.2, instead of the level of 7.55 it attained by 1913. In short, the difference in the age of German and British capital in steelmaking would have been about 3.6 years, not 5.0 years, after 1900, and less before. A 3.6-year age difference and a 3.0 percent growth in productivity would yield a cost difference attributable to slower growing British output of steel of less than 10.8 percent, substantially lower than Temin's estimate of 15 percent.

The second upward bias arises from the assumption that productivity change in steel was 3 percent per year when in fact, as was shown above, taking Bessemer and open hearth steel together it was only about 1.4 percent. The result of this revision is straightforward: 1.4 percent productivity change per year and 3.6 years of difference in age yields a productivity difference of only 5.0 percent, not 10.8 percent, if all productivity change was embodied.

The final source of upward bias, and the most serious, is the assumption that all productivity change was in fact embodied in new capital equipment. More precisely, it is implausible that the simple relationship between embodied productivity and gross investment, which is the central message of the argument, holds true. The general point is that in adjusting to the new lower rate of growth after 1890 the British steel industry would abandon the least productive investments first, keeping the highly productive ones.[21] That is, productivity would fall less than in proportion to the fall in the fresh capital built each year because those parts of the capital stock that were experiencing the most rapid productivity change and which therefore were most profitable to keep modern would be kept so. Although this general point is a conse-

21. The point was made by E. F. Denison in "The Unimportance of the Embodiment Question," *American Economic Review,* 54 (1964), 90–94, and by A. C. Harberger in "Taxation, Resource Allocation, and Welfare," pp. 25–70, in *The Role of Direct and Indirect Taxes in the Federal Revenue System* (Princeton: Princeton University Press, for the National Bureau of Economic Research and the Brookings Institution, 1964), pp. 66–68.

quence of profit-maximizing equilibrium, and might therefore be expected to be amenable to analysis, there is no obvious expression involving known parameters for the effect that changes in gross investment can in theory have on productivity.[22]

The alternative to a full theoretical treatment is to let the experience of the industry tell what the relationship actually was between the two variables.[23] The independent variable gross investment is observed with considerable error, and a regression of productivity change on it would have a slope biased towards zero. The crudest observations, however, suggest that changes in the growth rate of the industry had little effect on productivity. The growth rate of open hearth steel output in Britain, to take the most important case, fell from 18.3 percent per year in the 1880's to 7.0 percent per year in the 1890's, yet productivity growth slackened very little, if at all. The continued growth of productivity in the British hearth branch down to 1902 (recall that Bessemer productivity was everywhere stagnant) is very damaging to the embodiment argument, for by 1902 much of the adjustment to the new equilibrium age of capital had taken place and there was little further change possible in the British age relative to the American and German ages. As was shown earlier, by 1900 3.50 years of the difference of 3.65 years in 1913 had been achieved. If all productivity change after 1900 were embodied, Britain would have lost only 0.15 years worth of productivity from 1900 to 1913. At a rate of productivity change in the steel industry as a whole of 1.4 percent per year, the resulting difference be-

22. At any rate, not to my knowledge. The point involves in an essential fashion the existence of different kinds of capital goods with different rates of embodied productivity change. The problem is to find an expression for the effect of changes in the growth rate on the optimal accumulation of each kind of capital and on the resulting change in productivity.

23. This is essentially what Eitan Berglas did in "Investment and Technological Change," *Journal of Political Economy*, 73 (1965), 173–180. He was able to get a good fit to modern American manufacturing data for an equation that depends primarily on a time trend, i.e., disembodied technological change, and only weakly on gross investment.

tween British and foreign productivity would be about one-fifth of 1 percent.

There are some possible sources of downward bias in Temin's argument, but they are not enough to rescue the embodiment effect from insignificance, at least in its effect on relative British and foreign productivity in steel, as distinct from pig iron.[24] There is an extension of the argument, however, that one might hope could rescue some of its effect on the price of steel. A slower growing demand for steel would have slowed the growth of the pig iron industry and, in turn, the iron ore mining and cokemaking industries. The slower growing demand in these markets, through its effect on the average age of capital, would raise the price of the inputs of steel in Britain relative to inputs in America and Germany, raising the price of steel. In other words, the argument might go, the simple calculation is an understatement of the full effect of a relative slowing in the growth of British steel on the supply curve because it focuses exclusively on the first round of the effect.

The magnitude of the second round of the effect on the output price depends on the importance of the industry as a demander of inputs, the rate of embodied productivity change in these sup-

24. Although the pig iron industry is not mentioned by Temin, its growth rate was lower than steel's everywhere, while its rate of productivity growth in the United States was 0.75. In consequence, there may have been as much as a seven-year difference in the age of British compared with foreign capital and a 5.2 percent difference in costs. Because in 1913 there had been no sharp change for many years in the rate of growth of the American, German, and British pig iron industries, the asymptotic equilibrium ages of capital are adequate for comparisons. Between 1890 and 1913 the German industry grew at 4.14 percent per year, the American at 3.36 percent and the British at only 1.30 percent (see Chap. 3 above). The rest of the calculation assumes a 5-percent rate of depreciation and a 0.75 percent rate of productivity change. Nonetheless, a 5.2-percent productivity difference is not the order of magnitude of failure implied in the literature generally and in Temin's essay in particular. And this calculation does not meet the objection that there was not a relation of proportionality between gross investment and technological change.

111

plying industries, and their importance in steel costs.[25] If none of the inputs were specialized—that is, if steel consumed a negligible fraction of the total output of these industries—then the rates of growth of the supplying industries would be little changed by retardation and the second round of the effect on the supply curve would be small. Of course, the input of pig iron into steel was quite specialized, as were the inputs of ore and coke into pig iron. The pig iron industry and its suppliers, in fact, would have been the major source of second round effects on the price of steel. A crude estimate of the growth rate British pig iron would have achieved between 1890 and 1913 had steel output grown as fast as in Germany is about 3.7 percent per year, reducing the equilibrium age of capital about five years.[26] The age of capital in ore mining and cokemaking would probably have fallen by about the same amount. The consequent price falls in ore and coke would have reacted on the price of pig iron through the shares of ore and coke in pig iron costs (0.47 and 0.22) and the fall in pig iron costs on the price of steel through the share of pig iron in steel costs (0.30). Assuming embodied productivity change was 1.00 percent per year in ore and coke (which is a guess, but in all probability a high one) and 0.75 percent in pig iron, the total indirect effect of slower steel growth in Britain on the price of steel would have been approximately:

$$0.30 \ (0.75) \ (5 \text{ yrs.}) + (1.00) \ (5 \text{ yrs.}) \ (0.47)$$
$$+ \ (1.00) \ (5 \text{ yrs.}) \ (0.22) = 2.16 \text{ percent.}$$

25. The argument could be embodied in a general equilibrium model, perhaps an input-output model.

26. A German growth rate of 9.17 percent per year would have added 21.8 million tons of steel ingots to the 7.7 million tons actually produced by the British industry in 1913. If the proportions of open hearth and Bessemer steel were the same in this increment as in the actual production of 1913 and (allowing for scrap input and waste) if the open hearth and Bessemer processes required 0.8 and 1.1 tons of pig iron per ton of ingots, pig iron output would have been about 29.0 million tons rather than 10.3 million tons in 1913 (compared with 7.9 million tons in 1890). Assuming a rate of depreciation of 5 percent, the equilibrium ages are 16.4 and 11.5 years.

Correcting the estimate for the foreign retardation after 1900 and the incomplete attainment of equilibrium would reduce it well below 2.0 percent, which, again, is negligible. And, of course, these calculations are as vulnerable as the primary calculation to the objection that the embodiment effect is not proportional to the growth rate.

The steel industry is a likely place to find a significant embodiment effect. Certainly its growth rate in Britain dropped sharply after 1890, compared not only with the American and German steel industries but with British industry as a whole. Its links with other industries were strong. Contemporary and historical opinion has been nearly unanimous in believing that there was a technological gap to be explained, and many have supposed that the embodiment of technology in equipment could explain it. In a satisfying way, without recourse to assertions of entrepreneurial failure, the embodiment argument makes demand explain the alleged failures of supply. There are difficulties in applying this reasoning to the entire economy, to be sure, for a decrease in one demand frees resources to better supply another. Still, for the steel industry alone one might expect the explanation to have substantial force. It is apparent, however, that it does not. It does not provide an alternative to the hypothesis of failure in the years during which the rate of productivity growth in some sectors of the British industry was slower than it was abroad, even if one grants the assumption underlying both hypotheses that there was in fact a gap in the level of productivity to be explained.

7 American and British Productivity before 1913

It is time to measure directly the alleged gap between British and American productivity. If American productivity in iron and steel was well above British productivity in the twenty years up to World War I, the case for an interpretation of British performance in this period as entrepreneurial failure, or for any other interpretation that assumes British inferiority, would be strong. The common assumption is that in iron and steel as in other industries there was a large gap—10 or 20 percent differences in output for a given input would not surprise readers of the literature of late Victorian failure. Yet, notwithstanding the central place of the assumption in the historical picture of the industry, the measures of the gap have been casual. Scattered cases of American firms acting vigorously and British firms acting slothfully, the number of new items of technological knowledge discovered outside Britain, or, at best, the productivities of labor and capital in this material-intensive industry have been the main sorts of evidence and, as might have been expected, these doubtful measures of performance have given false readings of the relative productivities of the British and American industries.

Levels of Productivity. The procedure for comparing the level of productivity is to extend the measures used earlier for comparisons over time to comparisons over space. In pig iron the most convenient measure is the average product of coke, adjusted to reflect total productivity.

The most important adjustment is for mere differences in the iron content of American and British iron ores. From 1880 to 1913 British coke productivity and iron content of ore were roughly constant at 0.81 tons of iron per ton of coke and 0.41 tons of iron per ton of ore. During the same period, American iron content was also constant, but at a higher level, about 0.52 tons of iron per ton of ore. Since a 1 percent increase in the iron content of ore produced a roughly equal percentage increase in coke productivity in the late nineteenth century,[1] with ores of the quality typical in America British coke productivity would have been constant at more than 0.90 tons of iron per ton of coke from 1880 to 1913. The estimate cannot be far off, considering that the two most important pig iron districts in Britain, South Wales (which used ore equal in iron content to American ore) and Cleveland (which used ore very much poorer than American), reached this level of coke productivity by the 1890's.

American coke productivities, however, were below 0.90 tons of iron per ton of coke until 1900 and definitely surpassed it only during the first World War, as the data from the censuses given in Table 7 indicate. Table 8 compares coke productivities in a sample of eighty American coke blast furnaces in 1888–89 with coke productivities in British districts in 1887 and yields much the same result. The last line for the American sample shows the average coke productivities of furnaces that used ores with an iron content similar to those used in South Wales and Cumberland: The American furnaces had lower coke productivities. As in the census data, at similar levels of ore richness America had definitely lower coke productivity before 1900.[2]

1. This coefficient, 1, is roughly the average of the four coefficients used in Chap. 5 to correct coke productivity for changes in the iron content of ore in Britain.

2. The varying position of British relative to American coke productivity at equal iron content of ore was not owing to relative variations in the quality of coke in the two countries. Before the war, when American coke

TABLE 7. Average Coke and Ore Productivities in America.

Year	Tons of pig iron per ton of coke	Tons of pig iron per ton of ore
1879	0.70	0.51
1889	.83	.53
1899	.89	.52
1904	.88	—
1912	.92	.51
1917	.96	.51

Source: 1879–1904: *Census of Manufactures:* In order to make use of the data it was necessary to estimate coke equivalents of the bituminous and anthracite coal used in blast furnaces. The coke equivalent of bituminous coal was assumed to be reported productivities in the manufacture of coke from bituminous coal (approximately 0.63 tons of coke for each ton of coal). For anthracite, the assumed figure was higher to allow for the higher carbon content of anthracite (0.75). Because most American blast furnaces were working with coke alone by 1899, these rough adjustments significantly affect only the 1879 and 1889 estimates. The low level of coke productivity in these two years compared with later years is consistent with the rapid growth of total productivity reported in Chapter 6. "Ore," as in the British data, includes a small amount of scrap, cinder, and sinter.

1912 and 1917: American Iron and Steel Institute, *Statistical Reports.*

The business cycle between census years would not have affected coke productivity greatly, as a glance at the British experience (Fig. 3 in Chap. 5) shows. These numbers plus or minus 0.03 would surely give a band containing the true trend.

The appropriate inference would appear to be that for much of the period the British production function for pig iron, far from being below the American production function, was above it, and was at least equal to it down to World War I. There is one potential offset to this finding. If coke in America was cheap rela-

___
productivity was equal or inferior to British coke productivity, the quality of the coke used in the two countries was the same: in both countries the carbon and ash contents of the coke were roughly 88 and 8 percent.

TABLE 8. Coke and Ore Productivities in America (1888–89) and Britain (1887).

Location	Number of blast furnaces	Type of pig iron produced	Tons of pig iron per ton of coke	Tons of pig iron per ton of ore, cinder, and scrap
		America		
North	20	Bessemer	0.85	0.60
North	10	Forge and foundry	.81	.58
North	26	Mixed	.83	.59
South	24	Mixed	.69	.43
North (10) and South (1)[a]	11	Mixed	.80	.55
		Britain		
South Wales	All the region's blast furnaces	Largely Bessemer	.85	.53
Cumberland		Largely Bessemer	.86	.55
Cleveland		Mixed	.87	.35
All Great Britain		Mixed	.84	.44

Sources: United States: C. D. Wright, *Sixth Report of the Commissioner of Labor,* as in n. 10, this chapter; Great Britain: *Mineral Statistics of the United Kingdom* for 1887.

[a] Used ore with an iron content similar to that used in South Wales, selected from the above 80 blast furnaces.

tive to capital and labor, the American pig iron industry would have used relatively more coke than the British industry. In this case, American coke productivity would have been lower even if the American production function for pig iron was no lower than the British production function. Considering that British coal mining was facing sharply diminishing returns and that coal was 80 percent of the cost of coke, it is not surprising to find that coke was, in fact, cheaper in America relative to labor and capital than in Britain. The best statistics are for the 1880's. In 1889 a ton of coke was 40 percent less expensive in America relative to a man-year of labor than in Britain.[3] The relevant statistic for the comparison of coke and capital is $\dfrac{P_c}{(r + \delta)\, P_k}$. The interest rate, r, and the depreciation rate, δ, were decidedly higher in America than in Britain during the 1880's and the price of capital goods, P_k, was higher relative to the price of coke, P_c.[4] And since it was shown in Chapter 5 that British coke became steadily more expensive relative to capital and labor, it is likely that British coke continued after the 1880's to be relatively more expensive. British

3. American Connelsville coke sold at about $1.40 per long ton in 1889. In the same year, daily wages of northern pig iron workers were about $1.50. Assuming 310 working days a year, a man-year of labor cost $460. $1.50/$460 = 0.0033. Durham coke in 1889 sold at £0.42 per long ton. According to the British pig iron wage series (which in 1889 is close to its reliable base year, 1886), a man-year of labor cost £77. £0.42/£77 = 0.0055.

4. Comparing yields on safe bonds in the two countries in 1889, the American r was 4.5 percent and the British 2.8 percent. The linings of blast furnaces wore out in two and one-half years in America and in seven years in Britain, implying that δ was higher in America than in Britain. The yields on safe bonds are Macaulay's corporate unadjusted index number of yields of American railway bonds (*Historical Statistics,* p. 656) and the yield on consols (Mitchell, *Abstract of British Historical Statistics,* p. 455). The commentator on the American series remarks on railways that "for many years no other industry had as high a credit rating" (p. 651). The lining wear data are from Wright, *Sixth Annual Report,* p. 38, and Greville Jones, "A Description of Messrs. Bell Brothers' Blast Furnaces from 1844–1908," in the *Journal of the Iron and Steel Institute,* 1908, vol. 3, p. 59. The American comparison of P_k with P_c is for 1880–1882 and is based on detailed costs of construction of blast furnaces in E. C. Potter, "The South Chicago Works of the North Chicago Rolling-Mill Company," *Journal of the Iron and Steel Institute,* 1887, vol. 1, pp. 163–202. The British comparison is for 1887 and is based on B. Samuelson, "Notes on the Construction and Cost of Blast Furnaces in the Cleveland District in 1887," *Journal of the Iron and Steel Institute,* 1887, vol. 1, pp. 91–119. The cost of coke was compared with the cost of bricks, labor, iron hardware, blowing engines, boilers, and hoist engines.

ironmasters would have substituted cheap capital and labor for expensive coke, raising the average product of coke. In consequence, even if Britain and the United States had the same technology, British coke productivity could have been higher. That is, if labor and capital were good substitutes for coke, it may have been the case that America surpassed Britain in the making of pig iron earlier than the inflated British coke productivities imply.

The force of this objection, however, depends on the elasticities of substitution of labor and capital for coke, and these were probably low. The close agreement of the British total productivity measure with the average product of coke observed in Chapter 5 suggests this, as does the narrow range of technical substitutability apparent in the processes of the blast furnace. And even high elasticities of substitution would not imply American superiority.[5] The weight of evidence, in short, is that there was no substantial American superiority in pig iron before World War I. That is, the continued productivity growth in America after the 1880's, while productivity in Britain stagnated, was indicative of an initial American inferiority. Productivity in the American industry continued growing because it was catching up to the high average

5. The production function exhibiting constant returns to scale and neutral productivity change in coke (C), labor (L), and capital (K) can be written in the form:

$$Q/C = AF(K/C, L/C).$$

Taking the proportional change of both sides yields:

$$(Q/C)^* = A^* + S_K (K/C)^* + S_L (L/C)^*.$$

Now, $(K/C)^* = E_{KC} (P_c/r)^*$, where E_{KC} is the partial elasticity of substitution between capital and coke. Inserting this equivalence for both capital and labor into the expression yields:

$$(Q/C)^* = A^* + S_K E_{KC} (P_c/r)^* + S_L E_{LC} (P_c/w)^*.$$

The share of capital and labor together was about 0.25 in pig iron. Using high estimates of the proportional difference in price of 0.5 and elasticities of substitution of 0.5, the proportional British superiority in coke productivity (around 10 percentage points in the 1890's) would be reduced only 6 percentage points.

119

standard of the British industry, not because there had been a British entrepreneurial failure.

The same argument can be made for rolled steel (and, strictly speaking, rolled iron as well: but at the time the comparison is made rolled iron was a small part of the total). Here too it can be shown that American and British productivities just before the war were roughly equal. As was pointed out in Chapter 5 there is no simple physical measure of productivity in steelmaking (such as the average product of coke in ironmaking) and the comparison of productivities must rest on price data. The comparison here is productivity at home with that abroad rather than productivity in one year with that in another, but the principle is the same: the prices of the major inputs, labor and pig iron, are measured against the prices of the outputs of steel rails, plates, bars, sheets, and the rest.

The best sources for making uniform comparisons of the price structures of the two industries before World War I are the British Census of Production of 1907 and the American Census of Manufacturing of 1909.[6] Market prices reported in the trade journals are useful checks, but the census data has the advantage of broad coverage. Broad coverage has drawbacks, too, often merely concealing rather than curing the problem of heterogeneity in such categories as "American pig iron" or "British wage earners." Market prices, in contrast, refer to a specific commodity at a specific location sold under specific terms, although it is often difficult to determine what exactly these specifications were. The average values in the censuses, then, must be handled with care.

6. The years are calendar years. *Final Report of the First Census of Production of the U.K.,* p. iii: "The Census was taken in the year 1908 in respect of the year 1907, and . . . it is believed that the returns received are in the aggregate substantially representative of the production of the year 1907." U.S. Department of Commerce, Bureau of the Census, *Thirteenth Census of the U.S.,* vol. X, Manufactures, 1909 (Washington, D.C.: Government Printing Office, 1913), p. 1: "The statistics were collected during the year 1910, but relate in general to the year ending December 31, 1909."

The British census, for example, gives values and quantities for "thick plates," which include ship plates and boiler plates, while the American census gives data for "plates and sheets," which include sheets as well: since sheets, being thinner, required more rolling than plates, they were more expensive, and the effect of raising average values must be removed if the American industry is not to appear spuriously inefficient.[7] A similar problem of heterogeneity exists in the measurement of pig iron input. Low phosphorous (or "acid," "hematite," or "Bessemer") pig iron was expensive because it was cheaper to use and more expensive to make than other varieties. Because the British industry used more of it than the American industry the categories given in the censuses must again be broken down into more detail to avoid spurious productivity differences.[8]

The results of the various corrections are exhibited in Table 9, which gives the names and average values of comparable American and British inputs and outputs. The values that it was necessary to estimate are bracketed.

The marginal product of pig iron implied by this table—that is, the ratio of the price of pig iron to the price of a steel product such as structural shapes—is invariably higher in Britain. This is not surprising, since labor was, of course, much cheaper relative to pig iron in Britain than in America, causing labor to be used

7. The average value of plates and sheets is given in the American census as $40.00 per long ton (p. 240). The census gives tonnages, but not values, of plates and sheets by gauge (p. 238). Assuming that a gauge of 17 or lighter is comparable to the British category of "sheets" and assuming further that sheets sold at the black plate and sheet price (a high estimate of the true price) of $50.00 per ton (given at p. 240), the heavy plate price can be estimated at $36.10 per ton.

8. The object was to break the American census value of pig iron used in steelworks and rolling mills ($15.10 per ton, p. 252) into Bessemer and non-Bessemer prices, comparable to the values reported in the British census. The census gives tonnages for five types of pig iron; the American Iron and Steel Institute, *Report* for 1913 gives 1909 market prices for pig iron similar to each of the five types; from this information and the average census value it was possible to infer the appropriate census values for the Bessemer type and for the rest.

TABLE 9. Census Values per Unit of Pig Iron and Labor Inputs and of Major Outputs in British (1907) and American (1909) Rolled Iron and Steel.

American name	American value (dollars per ton or per man-year)	British name	British value (shillings per ton or per man-year)
Bessemer pig iron	[15.70]	Hematite pig iron	72.5
Other pig iron	[14.68]	Other pig iron	55.4
Average, all pig used	15.10	Average, all pig used	[65.3]
Yearly earnings of labor	679.00	Yearly earnings of labor	[1690.0]
Plates less than 17 gauge	[36.10]	Plates greater than or equal to ⅛"	138.7
Rails	28.38	Rails	120.0
Bars and rods	31.97	Steel bars and angles	133.8
Structural shapes	30.90	Girders, beams	128.0
Black plates and sheets	49.00	Black plates and sheets	188.0

Source: The source for America is the 1909 Census of Manufacturing, pp. 238–240 and for Britain the 1907 Census of Production pp. 101–103. The coverage is roughly half of the value of rolled products of iron and steel (i.e., excluding cast iron) in both Britain and America. The bracketed figures were estimated as described in previous footnotes. The British average value of pig used was estimated from the average values given in the 1907 Census (for Forge and Foundry, Hematite, and Basic) weighted by estimates of the hematite and nonhematite pig iron used to make steel and wrought iron (i.e., excluding cast iron and exported pig iron). These estimates, in turn, were based on the acid/basic proportions of steel output and an assumption that all puddling was done with nonhematite pig.

intensively and raising the marginal product of pig iron. What is surprising, in view of the usual assumption of overwhelming American superiority, is that this higher marginal product of pig iron in Britain is not outweighed by a correspondingly lower marginal product of labor. That is, as shown in Table 10, the uniformly high price ratio of pig iron to steel in Britain was just barely offset by a uniformly low price ratio of labor to pig iron, leaving very small differences in the total productivity of the American and British industries:

TABLE 10. Productivity Differences between Britain (1907) and America (1909) for Major Iron and Steel Products.

Products	Marginal products of pig iron (tons output per ton iron)		Percent difference in marginal products of pig iron (USA higher+)	Percent difference in total productivity (USA higher+)
	America (1)	Britain (2)	(3)	(4)
Heavy plates	0.418	0.471	−11.92	−1.57
Rails	.533	.545	−2.22	+8.13
Bars, rods, etc.	.473	.488	−3.13	+7.22
Structural shapes	.488	.510	−4.41	+5.94
Black plates and sheets	.308	.348	−12.20	−1.85

Source: Table 9 above. Col. 1 and 2 are the average price of pig iron divided by the price of the product. Col. 3 is [(1)−(2)]/[½ (1)+½ (2)]. Col. 4 is col. 3 minus the share of labor (assumed to be 0.192, which is the American value and is high for the U.K.) times the percentage difference in the price ratio of labor to pig iron (w/P_I). That is, col. 4 is

$$A^* = (P_I/P)^* + S_L (w/P_I)^*.$$

This can be shown to be identical to the A^* used earlier. The result assumes that labor and pig iron are the only two factors of production. In the United States they accounted for about 68 percent of the costs of production, excluding steel inputs into steel in the total cost.

On this reckoning, which agrees with the pattern of comparative advantage that might be expected, the American industry was slightly superior in the making of rails, bars, and structural shapes,

123

and the British industry in the making of plates and sheets. The assumption that the share of labor was the same for each product when it was in fact different imparts a small bias to the results, as can be seen from the steady relative improvement of British performance as the degree of fabrication (and labor-intensity) increases from rails to sheets. If this bias were corrected the productivity differences would be more uniform from product to product, with about 2 or 3 percent average superiority for the American industry.

A measured difference of 2 or 3 percent less output for a given input suggests that entrepreneurial failure had little to do with the relative positions of the British and American supply curves, especially in view of certain biases against the British industry in the measure. Two years of productivity growth, for example, between the year of the British census, 1907, and that of the American, 1909, would account for some of the difference. Moreover, 1907, and 1909 in the two countries were at different stages of their business cycles. The measures of productivity assume that the industries were in long-run equilibrium, that is, that their cost curves were flat. When a cyclical increase in demand has raised output to a new peak, straining capacity, this assumption is violated and productivity appears lower than it is in fact. And it is likely that the British industry suffered more from this downward bias in the measure.[9] In any case, though the measures in steelmaking, as in ironmaking, are crude, the result is plain: the total productivities of the American and British industries on the eve of World War I, when the allegations of American skill and British ineptitude were already many decades old, were virtually indistinguishable.

9. The year 1907 was a peak year for British steel output and was not surpassed until 1912. In 1909, the American industry produced about 3 percent more steel than in the previous peak, but the previous peak was three years before and was surpassed in 1910: the industry had ample time to adjust and was not pressed much beyond its previous peak output until 1910.

How Well British Entrepreneurs Performed. It is most doubtful, then, that the concept of entrepreneurial failure is helpful in understanding the iron and steel industry's history. The British industry faced slower growing demand than the American or German industries, especially after the substitution of steel for iron was completed at the end of the 1880's, but did not suffer a retardation of productivity growth on that account. The eventual cessation of productivity growth in the early 1880's in Bessemer steel, in the late 1880's in pig iron, and in the early 1900's in open hearth steel was a reflex of the exhaustion of available technology, not of slower growing demand. The rate of productivity growth in America did finally come to exceed the rate in Britain, but only because Britain had earlier achieved high levels of productivity. The good performance of British entrepreneurs is especially clear and significant in the open hearth. Here Britain was a leader in the process that was to become dominant in steelmaking everywhere; at a very early date, responding to the demands of shipbuilders for steel that would not crack, the industry had made it dominant in Britain. Indeed, the slow move to the open hearth in Germany and America could be spun into a tale of failure as readily as has the slow move to the basic process in Britain. Neither, of course, is justified. There was no failure in the neglect of basic ores and steel in Britain or the open hearth elsewhere: the rates of adoption of both were responses to the pattern of economic incentives.

In view of the competitiveness of the industry, the agile response of British entrepreneurs is hardly surprising. Less agile entrepreneurs would have paid a heavy penalty in the competitive milieu of the British industry. The potential rewards of agility, moreover, were great: so small was the share of capital in total costs that a 5 percent difference in productivity relative to his competitors would give an ambitious steelmaker 50 or 100 percent larger profits, for he, as the residual claimant, would reap as profit all the increase in output. Productivity differences even of this small magnitude, therefore, would not develop. It can be shown in fact that they did not. Considering the rapid productivity growth in all branches of the British industry in the 1870's and 1880's (except wrought iron, whose technology was stagnant everywhere), there is little doubt that in 1890 British productivity was at least equal to American. Supposing for the moment that it was merely equal (and not superior, as it was in fact), the known rates of productivity growth can be used to calculate the implied difference in the levels of British and American productivity in later years. By the early 1900's, on this reasoning, there could have developed a gap of only 3 percent at most in steel-making.[10] Down to about 1902, then, the case for important entrepreneurial failure is weak on even the most generous assumptions. After 1902 the case is better, for productivity growth in the open hearth branch ceased in Britain: by 1913 the difference could have been on the order of 18 percent, given the assumption of equal productivity in 1890. The assumption, however, is demonstrably false. There was nothing like an 18 percent difference in total productivity between the American and British industries before the war. The difference, in fact, was barely distinguishable from none at all. The inference is clear: the case for British entrepreneurial failure can not only be rejected, it can be reversed. Before World War I, with increasing certainty as successively earlier years are chosen for the comparison, a case can be made for the *superiority* of British over American productivity in iron and steel.

If the hypothesis of entrepreneurial failure has little merit, it may well be wondered why it has seemed so attractive to historians of the industry. The answer lies partly, perhaps, in an admirable desire to write conscientious, rather than apologetic, history. "If a business deteriorates," wrote Burnham and Hoskins in their *Iron*

10. To which could be added the gap attributable to pig iron alone. Productivity change in pig iron enters the measure of productivity for the iron and steel industry as a whole weighted by pig iron's share in total iron and steel costs (about one-third). In the 1890's American productivity in pig iron was growing at 0.75 percent per year and British was stagnating. One-third of 0.75 is 0.25 percent per year, or roughly 3.00 percent over the twelve years 1890–1902, except that the total rate of excess growth in the United States is 1.50 percent rather than only 0.25 percent.

and Steel in Britain, 1870–1930, "it is of no use blaming anyone except those at the top."[11] It is questionable whether an approach to history framed for the analysis of the careers of popes and emperors is appropriate for the analysis of ironmasters and steel merchants, but the impulse to a moral history cannot be faulted. The natural desire to affix blame, however, is liable to become an obstacle to understanding if it is combined with inadequate measures of performance. Economic history then descends to summary judgments of good and bad entrepreneurs the way the parody of schoolbook history chronicles Good and Bad Kings. The primary measure of performance for the industry has been its rate of growth. As was argued in detail in Chapter 3 and indirectly throughout, however, the rate of growth is inadequate as a measure of performance. Its use largely explains why contemporaries and historians have though it plausible that British entrepreneurs failed in the iron and steel industry and in the late Victorian economy as a whole.

After 1870 national income was growing much slower in Britain than in America and Germany, and the output of modern industry was growing slower still. Contemporaries were made aware of the contrast through statistics of exports and statistics of the output of a few industries, especially iron and steel, and were alerted to its apparently ominous implications by a growing army of critics. Historians have used more sophisticated statistics than they did, but the simple framework constructed by the critics for interpreting the statistics has lived on. Whatever its political and psychological significance, however, there was nothing economically ominous for Britain in the faster growth of two large, industrializing nations. It is unlikely that anyone should be blamed for Britain's failure to match their growth in any industry, least of all in an industry so dominated by internal supplies of resources and demands for investment goods as iron and steel. Late nineteenth-century entrepreneurs in iron and steel did not fail. By any cogent measure of performance, in fact, they did very well indeed.

11. P. 271.

Appendixes   Index

Appendix A Sources for Estimating U.K. Gross National Product and Inputs of Capital and Labor

1. *Gross National Product*

Gross national product is the sum of consumption, gross investment, government expenditure, and net sales of goods and services to the rest of the world.

a) *Consumption:* 1900 and 1910 are from D. A. Rowe's estimates, deflated to 1900 prices by him with price indexes of each item (given in Mitchell, *Abstract of British Historical Statistics* [henceforth Mitchell], p. 370). Estimates for 1870, 1880, and 1890 are from J. B. Jeffreys and D. Walters, "National Income and Expenditure of the United Kingdom, 1870–1952" in Simon Kuznets, ed., *Income and Wealth, Series V,* International Association for Research in Income and Wealth (1955), p. 27. The estimate for 1860 is from the extrapolation of the series of Jeffreys and Walters by Phyllis Deane and W. A. Cole, (*British Economic Growth, 1688–1959* [Cambridge, Cambridge University Press, 1964], p. 332). For 1860–1890, the estimates were deflated to 1900 prices by A. L. Bowley's index of the cost of living in *Wages and Income in the United Kingdom since 1860* (Cambridge,: Cambridge University Press, 1937), pp. 30–34. The 1860–1890 estimates of consumption were raised by about 2 percent to agree with Rowe in the period of overlap (1900–1913).

b) *Gross investment:* All years were from C. H. Feinstein,

"Income and Investment in the United Kingdom, 1856–1914," *Economic Journal,* 71 (1961), 374. He deflated the series to 1900 prices using a price index for each of ten component series.

c) *Government expenditure:* 1860, 1870, and 1880 are estimates in 1900 prices by J. Vererka, given at p. 37 by A. T. Peacock and J. Wiseman, *The Growth of Public Expenditure in the United Kingdom* (Princeton,: Princeton University Press, 1961). The estimates for 1890, 1900, and 1910 are in 1900 prices by Peacock and Wiseman, p. 42.

d) *Domestic commodity exports:* All years are from Albert Imlah's estimates, given in Mitchell, p. 283, deflated to 1900 prices using Imlah's price index of commodity exports, ibid., p. 331.

e) *Net exports of invisible:* All years are from Imlah, given in Mitchell, p. 333, deflated by L. Isserlis's index of tramp shipping rates given in Mitchell, p. 224. The year 1860 was extrapolated on the basis of export prices.

f) *Commodity imports, net of reexports:* The source is identical to that of commodity exports.

2. *Capital Input*

The capital series is a decumulation in 1900 prices from part of J. C. Stamp's capital stock estimate for 1914 by part of Feinstein's net investment series and an estimate of inventory investment.

a) *Capital stock in 1912:* The selection from Stamp's estimate (*British Incomes and Property* [London: P. S. King, 1916], p. 404) is that of E. H. Phelps-Brown and S. J. Handfield-Jones, "The Climacteric of the 1890's," p. 302, except that the value of revenue-yielding property of local authorities, estimated from loans outstanding (or, with the same result, a rough accumulation of the corresponding part of Feinstein's net expenditure by local authorities out of loans) as £219 millions (see A. K. Cairncross, *Home and Foreign Investment, 1870–1913* [Cambridge,: Cambridge University Press, 1953], p. 144), was deleted. Phelps-Brown and Handfield-Jones argue convincingly that Stamp's "1914" estimate is more appropriately attributed to 1912. According to Feinstein's implicit index of capital goods prices, values in 1912 can be taken to be the same as in 1900.

b) *Net investment:* Feinstein's series of net investment in 1900 prices was used, except that local authorities' loan expenditure was removed to avoid including investment in hospitals, schools, and so on, none of which appear in Stamp's estimate.

c) *Net inventory investment:* These were taken as 0.4 multiplied by the change in real gross national product. This procedure is defended in Phelps-Brown and Handfield-Jones, "The Climacteric of the 1890's," p. 304.

3. *Labor Input*

The labor series is the estimate of Phelps-Brown and Sheila Hopkins of total occupied population in the United Kingdom, given in Phelps-Brown and Handfield-Jones, "The Climacteric of the 1890's," p. 298. It refers to the census years 1861, 1871, and so on.

4. *Shares of Labor and Capital in Income*

The share of labor is 0.52, derived from the share of wages and salaries in home produced income estimated by Deane and Cole (*British Economic Growth,* p. 247). The share of capital (0.44) is a residual from 1.00 of labor's share and land's share (the latter based on Stamp's estimate of net property income). Land was assumed to be unchanging in quantity over the period.

Appendix B   The Prices of British Iron

and Steel

Only prices that are important to the argument and that are not readily available elsewhere are reported here.[1] The chief source, as it is for prices reported elsewhere in the literature, is the weekly *Iron and Coal Trades' Review* (henceforth ICTR), which published prices from the late 1860's onward. The usual procedure below for each year is to average the prices for the week of the ICTR closest to the middle of the months January, April, July, and October.

1. Mitchell (*Abstract of British Historical Statistics,* p. 493) gives prices of Scottish pig iron and Cleveland No. 3 pig iron and common bars for this period. Carr and Taplin (*History of the British Steel Industry, passim*) give many prices, usually from the ICTR, and Burn (*Economic History of Steelmaking*) gives some. These are the chief secondary sources of data on prices.

TABLE 11. Prices in the Iron and Steel Industry, 1864–1913.
(in shillings per long ton, except coke in pence per long ton)

	Cumber-land hematite pig iron	N. York-shire blast furnace coke	N. York-shire steel rails	Scotland steel ship plates	N. York-shire steel ship plates	N. York-shire iron ship plates
	(1)	(2)	(3)	(4)	(5)	(6)
1864	80.0 *s*	—	350.0 *s*	—	—	—
1865	—	—	340.	—	—	—
1866	—	—	—	380.0 *s*	—	—
1867	—	—	258.	—	—	—
1868	—	—	—	—	—	—
1869	—	—	220.	—	—	—
1870	72.5	121.0 *d*	210.	360.	—	—
1871	90.7	126.	239.	—	—	162.0 *s*
1872	142.5	192.	279	—	—	—
1873	169.2	410.	308.	—	—	272.
1874	109.6	312.	238.	—	—	222.
1875	81.6	186.	182.	—	—	177.
1876	74.1	144.	155.	—	—	154.
1877	72.8	131.	136.	302.	—	142.
1878	60.0	126.	113.	258.	—	122.
1879	57.5	114.	98.	228.	—	116.
1880	[70.0][a]	126.	118.	254.	—	158.
1881	61.1	114.	113.	222.	—	130.
1882	58.5	120.	111.	221.	—	139.
1883	51.8	126.	99.	200.	—	127.
1884	46.9	108.	90.8	175.	—	111.
1885	44.5	104.	95.0	144.	140.0 *s*	99.
1886	43.6	97.	83.8	139.	130.	92.
1887	46.1	97.	85.8	136.	126.	95.8
1888	44.7	97.	79.8	129.	138.	102.0
1889	52.2	173.	95.5	156.	146.	127.
1890	59.9[b]	140.	117.2	165.	151.	135.

135

Table 11 (*continued*)

	Cumber-land hematite pig iron	N. York-shire blast furnace coke	N. York-shire steel rails	Scotland steel ship plates	N. York-shire steel ship plates	N. York-shire iron ship plates
	(1)	(2)	(3)	(4)	(5)	(6)
1891	51.7[b]	140.	92.0	132.	124.	114.
1892	49.6	130.	93.5	124.	116.	107.
1893	46.0	130.	80.0	112.	104.	96.
1894	45.5	160.	75.	114.	102.	98.8
1895	46.5	150.	77.	105.	98.	96.8
1896	49.2	120.	93.5	108.	104.	98.8
1897	50.5	120.	91.5	114.	107.	104.0
1898	54.6	130.	90.8	118.	116.	—
1899	70.2	160.	108.0	146.	146.	—
1900	82.1	336.	148.	167.	164.	—
1901	61.0	180.	112.	124.	126.	—
1902	60.4	216.	110.	118.	115.	—
1903	58.6	216.	108.	115.	114.	—
1904	54.5	168.	93.	113.	111.	—
1905	61.1	192.	102.	124.	122.	—
1906	69.8	204.	125.	146.	140.	—
1907	78.2	252.	134.	152.	150.	—
1908	60.7	192.	118.	127.	124.	—
1909	59.6	180.	108.	123.	119.	—
1910	67.1	204.	110.	128.	131.	—
1911	64.9	168.	114.	138.	136.	—
1912	75.8	240.	123.	152.	155.	—
1913	78.1	240.	134.	164.	160.	—

Sources:
Col. 1. *Cumberland Hematite Pig Iron* (shillings per long ton).
Cumberland prices for hematite (that is, low phosphorous or Bessemer) pig iron are available from an early date because the Cumberland region was the main domestic source of nonphosphoric ore. During the 1880's the ICTR starts mentioning pig irons

made in North Yorkshire and Scotland (from imported ores) that were also appropriate for acid steelmaking. The detailed sources for the Cumberland series are:

1864: Bell, *The Iron Trade of the United Kingdom* (London: British Iron Trades Association, 1886), p. 20, "hematite pig."

1870–1872: No. 3 Bessemer pig in Northwest England, ICTR, four dates per year averaged.

1873–1913: Same definition, ultimately from ICTR, yearly averages, given in *Mineral Statistics* down to 1913, Sessional Papers, 1914–1916, vol. 80, p. 272.

Col. 2. *North Yorkshire Blast Furnace Coke* (pence per long ton).

Cleveland and Durham were the major sources of coking coal. The reporting of prices is best for these districts, although it still requires supplementation. Notice that the prices are often in multiples of 12 or 6: this is because the price in the source was rounded to the nearest shilling or sixpence. The third digit, therefore, is in most years of doubtful significance.

1870–1883: Durham coke at ovens of one maker, Bell, *Iron Trade,* p. 30.

1884, 1888–1892, 1900–1913: Cleveland furnace coke at ovens, ICTR reports on the Cleveland trade and tables (in later years), one spring quotation per year.

1885–1887, 1893–1899: Extrapolated on the basis of the U.K. export price of coal. The rounding to two significant digits is clearly appropriate.

Col. 3. *North Yorkshire Steel Rails* (shillings per long ton).

"North Yorkshire" is a mislabeling before 1880.

1864: Bell, *Iron Trade,* p. 20, probably a Sheffield price, as most of the prices before 1880 were: there was no steel made in North Yorkshire at this early date.

1865: Alan Birch, *The Economic History of the British Iron and Steel Industry, 1784–1879* (London: Frank Cass, 1967), p. 354, a Sheffield price.

1867: Ibid., p. 359, a South Wales price.

1869: Bessemer's evidence to the Royal Commission on the Coal Supply, Sessional Papers, 1871, vol. 18, p. 490.

1870–1879: J. S. Jeans, *Steel: Its History, Manufacture, and Uses* (London: 1880), p. 705; "the average realized price of Bessemer steel rails at some of the largest works in England." They agree with the scattered price quotations in ICTR in the 1870's, being somewhat below them when prices fell, as might be expected. The 1879 price is given in the report of the British Iron Trade Association (henceforth BITA) for 1880, p. 41.

1880–1882: Prices are missing from the ICTR in these years.

Those given here were obtained by regressing the known prices in 1876–1879, 1883, and 1885–1888 on the declared value per ton of steel exported (given in BITA, 1890, p. 34). The fitted regression equation is $P_R = 19 + 0.70 P_X$ ($R^2 = 0.95$). The London price of rails, given in the ICTR for 1880 and 1881 moves with the estimated prices.

1883, 1885–1913: Heavy steel rails on the Northeast Coast, ICTR, four dates per year averaged.

1884: Extrapolated on the basis of the relationship with rail prices in Cumberland in the two adjacent years.

1910: Same procedure as 1884.

Col. 4. *Scotland Steel Ship Plates* (shillings per long ton).

Steel was used very seldom for ships before the late 1870's see Chap. 3, second section, above).

1866, 1870: BITA, 1880, p. 46, prices paid for *cast* steel (not open hearth or Bessemer) plates at one or two of the larger shipbuilding works in Britain. The source gives prices for more years, but these two illustrate the prevailing level.

1877–1882: Based on the graph of Scottish open hearth ship plate prices given in a paper by J. Riley in the *Transactions of the West of Scotland Iron and Steel Institute*. The graph was reproduced in several places, for example, in BITA, 1885, p. 141. Riley was the works manager of the Steel Company of Scotland, one of the first users of the ferromanganese open hearth process. During 1883–84 (eight quarterly observations) the price was 10 to 20 shillings below the ICTR quotations. The price was therefore raised according to the relationship that holds during 1883–84 ($P_{ICTR} = 53 + 0.82 P_{Riley}$). Other evidence (for example, BITA, 1880, p. 46) confirms the resulting series.

1883–1913: Ship plates in Scotland, ICTR, four dates per year averaged.

Col. 5. *North Yorkshire Steel Ship Plates* (shillings per long ton).

1885–1913: Steel ship plates on the Northeast Coast, ICTR, four dates per year averaged. For 1885 and 1886 the average had to be filled out in the January week by an extrapolation from the Scottish price. The later beginning of the North Yorkshire series in the ICTR is indicative of the somewhat slower adoption of steel there.

Col. 6. *North Yorkshire Iron Ship Plates* (shillings per long ton).

1871, 1873–1897: Manufactured (wrought) iron ship plates on the Northeast Coast, ICTR, four dates per year averaged. 1871 is for the first two quarters only and 1879 for the last two quarters only. The series is discontinued here in 1897, not because it could not be continued but because it is of little interest after the 1890's.

a 1880: The price of Bessemer pig iron given in the source is 84.5 shillings (Bell, *Iron Trade*), p. 21, gives 81.5s. This is unreasonably high by comparison with Cleveland rail prices or Scottish ship plate prices. The realized price of Cleveland No. 3 pig iron (phosphoric) increased 19 percent from 1879 to 1880, the price of Scottish pig iron 16 percent, the price of imported ores (largely nonphosphoric) 4 percent, and the price of Cleveland coke 10 percent. In view of these magnitudes of price change in its inputs and substitutes, the price of Bessemer pig iron in Cleveland and Scotland could not have increased 47 percent, as the 84.5 shilling price implies. The sharp rise of Cumberland ore prices in 1880 agrees better with it, but a plot of the Cumberland pig iron prices against the ore prices (the prices are given in J. D. Kendall, *The Iron Ores of Great Britain and Ireland* [London: 1893], pp. 392–393), suggests that 70 shillings is more credible. The 70 shilling price was used.

b 1890, 1891: Missing in the source. Extrapolated from the ICTR's prices of mixed numbers of Northwest Coast Bessemer pig iron, four dates per year averaged.

Appendix C The Input and Output Structure of the British Iron and Steel Industry in 1907

In 1908 the Board of Trade began the collection of questionnaires on the activities of nonservice industries in 1907 for Britain's first, timid census of production. The census legally required no information on wages, the quantities of materials used, or the value of the capital stock, no breakdown of the values of materials by kind, no breakdown of the value and quantity of output in an industry any finer than that in the export and import lists, and no statement of quantities of output for any commodity not on the lists. In the iron and steel industry, however, some of the missing information was supplied in response to voluntary census questions, and the rest can be inferred from other sources to give a reasonably full and accurate picture of the industry before World War I.

Each of the three processes in the making of iron and steel products has its characteristic equipment. The first is the blast furnace, in which oxygen and some other impurities are removed from the iron ore by burning it with coke and limestone. The product, pig iron, acquires carbon in this process, leaving it brittle and unsuitable for most uses. Of the £33.2 million worth of pig iron produced in 1907, about £7.2 million went into exports, about £5.8 million into cast iron, and the remainder into the second piece of equipment, the decarburizing furnace. By 1907 the older decarburizing process, puddling, took only about £4.0

million worth of pig iron, the rest (£16.2 million) going to the newer Bessemer and open hearth steel processes. The products of these puddling and steelmaking furnaces are low in carbon and therefore malleable: they can be shaped in the final characteristic piece of equipment, the rolling mill, emerging as rails, plates, bars, sheets, and other products. The census data makes it possible to trace in adequate detail the flow from ore mine to final product in 1907. Table 12 is read as follows. Reading across one row, say the row marked "Pig iron," gives the values of the product (in this case the product of the pig iron branch of the industry) flowing into other branches of the industry (cast iron using £5.78 million worth of pig iron, puddled iron using £4.05 million, and so forth, through exporters using £7.19 million). Reading down one column gives the value of inputs (pig iron, scrap, value-added, and so forth) used by the purchaser named (pig iron, cast iron, and so forth, through final domestic purchasers of the products of the iron and steel industry). Note that the inputs purchased from outside the iron and steel industry (coal, miscellaneous purchases, value-added, and imports) could not be allocated among the branches of the industry at a later stage of production than the making of pig iron because of a lack of detailed data on the costs of production in these stages.

The italicized figures in the last line are the gross outputs of the various products: when combined with the £7.19 million exports of pig iron they constitute the £85 million sales of the heavy iron and steel industry to other industries. The finished outputs (that is, excluding pig iron) are classified in the table by the form of iron and steel used in their manufacture. In 1907 the traditional cast and wrought iron materials were, surprisingly, still used in over a quarter of the finished output by value. Bessemer steel, which is the material that receives the most attention in the historical literature, was used in only one-fifth of the output and even at its height of relative importance in the late 1880's could not have been used in much more than a quarter. The dominant material in 1907 was open hearth steel: over half of the value of finished output of the industry was made from it.

TABLE 12. The Input-Output Relations of British Iron and Steel in 1907. (value, £ millions)

Sales from: \ Sales to:	Pig	Cast	Puddled	Rolled wrought	OH	Bess.	Rolled steel	Final purchasers Export	Final purchasers Domestic
Pig iron	0	5.78	4.05	0	9.65	6.37	0	7.19	0
Cast iron	0	0	0	0	0	0	0	1.94	13.11
Puddled iron	0	0	0	5.10	0	0	0	0	0
Rolled wrought iron	0	0	0	0	0	0	0	1.69	5.06
Open hearth steel	0	0	0	0	0	0	21.52	0	0
Bessemer steel	0	0	0	0	0	0	8.22	0	0
Rolled steel	0	0	0	0	0	0	0	20.59	35.98
Scrap	0	0	0	0	5.19	0	0	0	0
Ore	14.38	0	0	0	0	0	0	0	0
Coal (coke in pig iron)	11.63			3.84				—	—
Miscellaneous purchases from other industries	0.72			23.76				—	—
Value-added	6.46			23.58				—	—
Imports				2.41				—	—
Gross output (total)	33.19	15.04	5.10	6.75	21.52	8.22	56.57	31.41	54.15

Source: The gross outputs of final products are the sums of values given in the census. Intermediate outputs, such as pig iron or Bessemer steel ingots, were estimated on the basis of voluntary statements of values per ton given in the census, as were, for the most part, the inputs of scrap, ore, and coal. Both coke and ore are estimated at their cost at the blast furnace, i.e., including transport. The value-added for all iron and steel is given in the census; the value-added in pig iron alone is derived as a residual from the total value of pig iron. The value-added in the later stages of production, then, is a double residual: some of the miscellaneous values undoubtedly belongs in value-added.

142

Appendix D   Sources and Methods for Productivity Measurement in Pig Iron

The measure of total productivity change depicted in Figure 4 in Chapter 5 is calculated as:

$$A^* = Q^* - S_Q Q^* - S_C C^* - S_L L^* - S_K K^*.$$

There is a range of ambiguity in the measure between two years because it varies with the year chosen as base. It can be shown that mixing year 1 shares with year 2 quantity-changes provides one bound on the true measure and the reverse mixture the other bound. The range of ambiguity was small enough to ignore. Figure 4, then, is the arithmetic average of the bounds linked by successive multiplication to an arbitrary base of about 1.2 in 1879.

The sources can be described under the headings of each element in the productivity formula. The ultimate sources of much of the data were the annual volumes of *Mineral Statistics of the United Kingdom.* They report pig iron output, ore (imported and domestic) and coke inputs, the value of domestic ore at the mines and imported ore at the ports, and the number of furnaces in blast. The other data come from a variety of sources.

1. *The Quantity of Pig Iron*

Mineral Statistics, as in Mitchell, *Abstract of British Historical Statistics,* pp. 131–132 (incidentally, the totals for 1885 and 1886

in Mitchell should be reversed: 1885 = 7415 and 1886 = 7040).
Mineral Statistics gives a more detailed regional breakdown than
Mitchell.

2. *The Price of Pig Iron*

The price times the quantity is the total gross revenue, required
to calculate cost shares. In order to avoid misstatements of gross
revenues (which would greatly distort the share of capital series,
since capital's share is defined as a residual), as detailed price
statistics as possible were used. The primary quality difference is
between the cheap foundry pig iron (for example, Cleveland
No. 3) and the expensive hematite or "Bessemer" pig (for example, South Wales pig made from Spanish rubio, Cumberland
pig made from domestic hematite ores). Therefore, before 1890,
when the quantity of hematite production was first reported
separately, Cumberland hematite pig prices were used to value
Cumberland, Lancashire, North and South Wales pig output, and
Cleveland pig prices for what was left. Cleveland also produced
some hematite iron in this period, but it cannot be separated even
approximately from the foundry and basic iron. Scottish iron was
valued separately as well, since it was always considerably more
expensive than Cleveland pig and separate statistics were collected
for it. The series of Cumberland hematite prices was taken from
Carr and Taplin, *History of the British Steel Industry, passim*
(their prices were collected from the weekly price lists in the *Iron
and Coal Trades' Review*).

3. *The Quantity of Ore*

The quantity of domestic ore *mined* and foreign ore *imported*
was used (ultimately from *Mineral Statistics*). A more correct
concept would be the quantity of ore actually consumed at blast
furnaces, rather than ore mined and imported, which is one market
step removed from the blast furnace. Although the consumption
figure is available in *Mineral Statistics* (for example, 1873–1914

144

in *Mineral Statistics* for 1914, Sessional Papers [henceforth S. P.], 1914–1916, vol. 80, p. 270), it was not used. In every year except 1882, 1885, and 1887, consumption is higher than the sum of imports and mined production, probably because it included mill and forge cinder as well as ore. The bias from the use of the ore production figure is unimportant for measuring the long-run course of productivity change.

4. *The Share of Ore in Total Costs*

Mineral Statistics gives the total value of domestic ore at the mines and the total value of imported ores at the point of importation. Some correction must be made for transport costs from the mines or ports to the blast furnaces. A full estimate of transport costs based on detailed estimates of the movement of materials and the cost of movement was not attempted. A few scattered estimates indicate that transportation cost about 20 percent of the value of the ore at the mines and ports. For example, the Committee on Industry and Trade reported that the rail transport costs of ore for three Cleveland firms were about 6.7 percent of the blast furnace price of the pig iron in 1914 and 7.9 percent in 1925 (*Factors in Industrial and Commercial Effiicency* [London: HMSO, 1927], vol. 1, p. 518–520). Since ore values were about 38 percent of pig iron values (there was considerable variation from year to year), the Committee figures imply transport costs of 18–21 percent of ore values. Such estimates of transport costs are crude, but errors in them do not greatly affect the results of the productivity measurement. A rough numerical example will show why this is so. Suppose transport costs of ore were 30 percent rather than 20 percent of the value of ore at the mines and ports. The effect of the increase would be to increase the share of ore costs in total costs and reduce the share of the residual factor, capital. The ore share would be increased $(0.38)(0.10) = 0.038$ and capital's share correspondingly reduced. From 1870 to 1913 the amount of ore increased by an average of about 0.60 percent per year and capital about 0.08 percent per year, implying a reduction in the average growth of productivity owing to the higher transport cost adjustment by 0.02 percent per year, which is negligible.

5. The Quantity of Coke Used

Like the ore series, the ultimate source for the coke used was *Mineral Statistics*. The average product of coke in each region of pig iron-making follow much the same pattern as the average product in the entire country.

6. The Share of Coke in Total Cost

The procedure was to multiply the quantity of coke by a price of coke at the ovens; dividing the product (the value of coke used) by the total value of pig iron output in each year gives the share of coke. There is no correction for transport costs. The transport burden of coke was generally not as high as that of ore because coke ovens were frequently at the blast furnaces (a coke price at the ovens, then, would include the transport costs of the coal used to make coke). In any cast, the situation is analogous to correcting iron ore values for transport and even a fairly large error can be accommodated without affecting the results. The sources of the coke price is in most years a spring issue of the *Iron and Coal Trades' Review*. The *Review* gave coke prices at the Durham ovens irregularly. For the years from 1870 to 1883, I. L. Bell's series in *The Iron Trade of the United Kingdom*, p. 30, was used. In the years 1885–1888, 1893–1894, and 1897–1899 the coke price had to be extrapolated from the export price of coal. Throughout the period, Durham and North Yorkshire produced a large share of the coke made in the country. If a single price must stand for the whole, the Durham price is the best.

7. The Employment of Labor at Blast Furnaces

Before 1894 there is no useful labor input series. Therefore, the labor series was extrapolated back to 1870 from 1894 on the basis of a constant average product and share of labor. The share is 10 percent. Before 1894, then, labor exerts no influence on the measure except to reduce the proportional change in output by 10 percent. After and including 1894 the January returns in the *Ministry of Labour Gazette* of "number of workpeople at blast furnaces" were used, inflated by 11 percent to allow for the representation in the survey of about 90 percent of the furnaces in blast. The series is only accurate to within ± 10 percent.

8. *The Share of Labor in Total Costs*

The 10 percent share of labor assumed before 1894 agrees fairly well with later shares. The two benchmark years for the national wage data are 1886 and 1906, the other years being interpolated from 1886 on the basis of G. T. Jones's series of the wage of slaggers at Cleveland blast furnaces (*Increasing Returns,* p. 279, col. 8: his hourly wage series was changed into a shift series—the shift was reduced from 12 to 8 hours in 1898). The 1886 figure is the yearly wage of £73 per year from the survey of the Royal Commission on Depression of 1886. Burn remarks of the next benchmark, £81 per year from the "Report of an . . . Enquiry into Earnings . . . in the Metal Industries . . . in 1906" (S.P. 1911, vol. 78, p. xvii), that it is "only useful in fixing an impossibly low limit" (Burn, *Economic History of Steelmaking,* p. 137). Burn appears to have been correct in this judgment, for the extrapolation from 1886 to 1906 on the basis of Jones's index is higher (namely, £89). The extrapolation has been used here.

9. *The Quantity of Capital Services*

The estimate is a rough measure in each year of the cubic volume of the furnaces in blast. The furnaces-in-blast component of the capital series was, as usual, from *Mineral Statistics.* It varies a great deal, suggesting that capital employment was flexible and, therefore, that underutilization will not distort the productivity series over the business cycle (this matters little for the purposes of Chapter 5, for it is concerned with long tendencies, not the business cycle). The other component of the capital series is an estimate of the cubic capacity of a typical furnace at any time. This estimate is based on a variety of scattered sources, including contemporary statements of typical blast furnace heights and photographs and diagrams of blast furnace lines: for example, A. H. Sexton and J. S. G. Primrose in *An Outline of the Metallurgy of Iron and Steel,* 2nd ed. (Manchester, Eng.: Scientific Publishing Co., 1907) give some current typical measurements of blast furnaces in Cleveland, Scotland, and South Wales; again, D. E. Roberts, "Blast Furnace Design, 1867–1927" and A. K. Reese, "A Review of Blast Furnace Practice, 1867–1927," both in the Diamond Jubilee Issue (December 1927) of the *Iron and Coal Trades' Review,* give rough estimates of blast furnace dimensions back to 1867. The physical volume of furnaces is attractive as a measure of capital services because it captures some of the services of the material-handling equipment, which is likely to be related to furnace volume. On the other hand, the cost of furnaces is largely the cost of the brickwork and the brickwork is proportional to the surface area of the furnace. It is shown in Chapter 5, however, that substituting surface area for cubic volume makes little difference in the productivity measure.

10. *The Share of Capital in Total Costs*

This is simply a residual from the other shares: $s_K = 1 - s_O - S_C - S_L$.

Index

Note: The first occurrence of each work cited in the footnotes is given here under the author's name.

Acid (hematite, Bessemer) steel. *See* Basic process

Aldcroft, D. H.: ed. *Development of British Industry and Foreign Competition,* 9n; "The Entrepreneur and the British Economy," 2, 3n; generalizes from iron and steel, 2

American Iron and Steel Institute: *Annual Statistical Report,* 40n

Arrow, K. J., and others: "Capital-Labor Substitution," 52n

Ashworth, W., *Economic History of England,* 8n

Barker, T. C.: "The Glass Industry," 9n

Basic process: description, 57; many believe neglected irrationally in U.K., 57–58; Burn believes irrational neglect of Lincolnshire ores, 59; Burn incorrect, 60–67; early technique requires very phosphoric ore, 59; later used wider variety of ores than other processes, 68; neglect of Lincolnshire ores cannot account for neglect of basic process, 68–69; rapid move to process in *1900*'s, 70; Talbot furnace makes process profitable in U.K., 71; Chapter 4, *passim*

Berglas, E.: "Investment and Technological Change," 110n

Bell, I. L.: *Iron Trade of the U.K.,* 50n; attributes slow adoption of steel to high price, 50

Bessemer process: dominates rails, 46, 99; early improvements, 47–48; not useful for ship plates, 47–50; most output for rails, 53; declining importance in U.K. after *1880*'s, 53, 141; important in U.S. much later, 53–54; only slight shift of advantage toward basic process in *1900*'s, 70; heroic age before *1870* exaggerated, 85; slow maturation, 89; royalty drops in *1870,* 90. *See also* Productivity, rate of change; Railmaking steel

Board of Trade, U.K.: *First (1907) Census of Production,* 16n; *Enquiry into Earnings . . . of Labour,* 75n

149

Index

Borenstein, I. *See* Creamer, D., and others

Bowley, A. L.: *Wages and Income,* 16n

British Iron and Steel Federation: *Statistics,* 40n

Bureau of the Census, U.S.: *Historical Statistics,* 42n

Burn, D.: *Economic History of Steelmaking,* 2n; sets tone for hypothesis of failure, 1, 6; difficult to measure enterpreneurship, 21; old firms and bad entrepreneurship, 35; small markets inhibit technological change, 36–37; neglect of by-product coking, 56n; neglect of basic steelmaking, 57, 58–60; neglect of basic steelmaking identified with neglect of Lincolnshire ores, 69; basic process forced on U.K. by inelastic supply of acid pig iron, 70. *See also* Basic process; Chapter 4 *passim*

Burnham, T. H., and Hoskins, G. O.: *Iron and Steel in Britain,* 2n; blame men at top, 2, 126–127; basic process and ores neglected, 57

Businessmen. *See* Entrepreneurial failure in Britain, historiography

Canals and Waterways, Royal Commission on, 63n

Cannan, E.: "The Practical Utility of Economic Science," 5n; military metaphors of economic defeat, 5

Capital stock: iron and steel not capital intensive, 75–76; unimportant to measure closely in pig iron, 147–148. *See also* Embodiment

Carr, J. C., and Taplin, W.: *History of the British Steel Industry,* 26n

Cartels. *See* Monopoly in British iron and steel

Census of Manufactures (American, *1879–1914*), 100

Census of Production (British), 16n

Chenery, H. B. *See* Arrow, K. J., and others

Church, R. A.: "The Effect of the American Export Invasion," 9n

Clapham, J. H.: *Economic History of Modern Britain,* 6n; widespread entrepreneurial failure, 6; output a bad index of success, 6, 40; conservativism in adopting steel in ships, 47; basic process neglected, 57; Athena model of productivity in Bessemer steel, 89–90

Climacteric in British productivity: related to failure in iron and steel from *1870*'s by Coppock, 2; dating in 1870's by Coppock supports hypothesis of failure, 7; named by Phelps-Brown and Handfield-Jones, 7; iron and steel too small to explain, 15–20; railways, chemicals, mining, and iron and steel together too small to explain, 20n; difference in timing in different parts of iron and steel belies single cause, 94. *See also* Entrepreneurial failure in Britain, historiography; Productivity, rate of change

Cole, A. H.: "An Approach to the Study of Entrepreneurship," 7n

Commissioner of Labor: *Sixth Annual Report,* 104n

Competition. *See* Monopoly in British iron and steel

Conrad, A. H., and Meyer, J. R.: "Economics of Slavery," 62n

Coppock, D. J.: "The Climacteric of the 1890's," 2n; climacteric related to failure in iron and steel from the *1870*'s, 2; early

150

dating of climacteric supports hypothesis of failure, 7. *See also* Climacteric in British productivity

Court, W. H. B.: *British Economic History,* 7n

Creamer, D., and others: *Capital in Manufacturing,* 98n

Crossley, D. W. *See* Pollard, S., and Crossley, D. W.

David, P. A.: "Landscape and the Machine," 11n; use of interrelatedness, 11

Deane, P. *See* Mitchell, B. R., and Deane, P.

Demand. *See* Railmaking steel; Shipbuilding steel; Exports; Embodiment; Market scale of British iron and steel; Chapter 3, *passim*

Denison, E. F.: "The Unimportance of the Embodiment Question," 109n

Dobrovolsky, S. P. *See* Creamer, D., and others

Domar, E.: "On the Measurement of Productivity Change," 18n

Dunlop, J. T.: "Price Flexibility and the 'Degree of Monopoly'," 25n

Earnings and Hours of Labour: Board of Trade *Enquiry into,* 75n

Embodiment: popular alternative to entrepreneurial failure, 10, 105, 113; applied to steel by Temin and others, 11, 105–106; theory developed, 106–107, 109–111; rejected as trivial in magnitude, 106–113

Entrepreneurial failure in Britain, historiography: generalizations from iron and steel, 1–3; full statement of hypothesis in *1960*'s

by Landes and others, 2–4, 7; list of charges, 4; criticism of hypothesis by historians, 6, 7–8; opinion in textbooks of *1960*'s, 7n; economists substitute alternative hypotheses, 9–11, 105, 113; inadequacies of alternatives, 11–12, 105–113; failure rejected in recent quantative studies, 12n; iron and steel best case for test, 14–15, 20

alleged failures in British iron and steel: family firms, 35–36; ignore economies of scale, 36–37; output growing slowly, 39–40; slow adoption of steel in ships, 46–47; slow adoption of by-product coking, 56–57; neglect of basic ores and basic steel process most important, 57–59, Chapter 4, *passim;* general technological gap, 73, 105, 113–114, Chapter 7 *passim*

reasons for the popularity of the hypothesis of failure: metaphor, 1, 5, 126–127; output as measure of success, 4–5, 6, 39–40, 44–45, 55, 127; narrative, 6, 8; mismeasurement of productivity, 7, 11–14, 114; hindsight identifying innovations of the future, 54, 56, 57; conscientious history, 126–127

Erickson, C.: *British Industrialists,* 25n

Exports: bad salesmanship, 4; rails chiefly for export, 34, 53; tariffs abroad make U.K. dependent on home market, 41–44; finished products exported chiefly rails, 142. *See also* Market scale of British iron and steel

Index

Family firm: significance questioned, 35–36
First Census of Production (U.K., *1907*), 16n
Fishlow, A.: *American Railways,* 19n
Fogel, R. W.: *Railroads and American Economic Growth,* 19n
Foreign trade. *See* Exports; Market scale of British iron and steel
Frankel, M.: "Obsolescence and Technological Change," 9n; theory of interrelatedness, 9

Gordon, D. F.: "Obsolescence and Technological Change," 9n
Giesen, W.: "The Special Steels," 50n
Giffen, R.: "miscellaneous industries and incorporeal functions" perform well, 8
Gilchrist, P. C.: manufacturers blamed for neglect of basic process, 58
Griliches, J. *See* Jorgenson, D. W., and Griliches, J.

Habakkuk, H. J.: *American and British Technology,* 2n; failure in iron and steel caused by slow growth, 2, 10, 105. *See also* Embodiment
Handfield-Jones, S. J. *See* Phelps-Brown, E. H., and Handfield-Jones, S. J.
Harberger, A. C.: "Taxation, Resource Allocation and Welfare," 109n
Hematite (acid, Bessemer) pig iron. *See* Basic process; Chapter 4, *passim*
Hobsbawm, E. J.: *Industry and Empire,* 8n
Hobson, J.: *Incentives in the New Industrial Order,* 6n
Hoffman, R. J. S.: *Great Britain*

and the German Trade Rivalry, 4n
Hoffman, W. G.: *Wachstum der deutschen Wirtschaft,* 14

Interrelatedness: alternative to entrepreneurial failure, 9; rejected except for agriculture, 11–12
Iron ore. *See* Ore, iron

Jeans, J. S.: *Iron Trade of Great Britain,* 58n; consumers blamed for neglect of basic process, 57–58; evidence on transport costs, 63, 64, 66
Jones, G.: "Description of Messrs. Bell Brothers' Blast Furnaces." 118n
Jones, G. T.: *Increasing Returns,* 83n; invented price measure of productivity change, 103–104; overstated productivity change in U.S. pig iron, 104
Jorgenson, D. W.: "The Embodiment Hypothesis," 86n
Jorgenson, D. W., and Griliches, Z.: "The Explanation of Productivity Change," 96n

Kendall, J. D.: *Iron Ores of Great Britain,* 67n
Kendrick, J. W.: *Productivity Trends in the United States,* 14
Kindleberger, C. P.: "Obsolescence and Technical Change," 10n; *Economic Growth in France and Britain,* 10n; institutional theory of interrelatedness, 9–10; old firms and bad entrepreneurship, 35
Koutny, E. *See* Reuss, C., and others

Landes, D. S.: "Technological Change and Development in

Western Europe," 3n; "Entrepreneurship in Advanced Industrial Countries," 3n; eloquent statement of hypothesis of failure, 3–4; generalizes from iron and steel, 3n; dates failure from *1870*'s, 14; old firms and bad entrepreneurship, 35. *See also* Entrepreneurial failure in Britain, historiography
Lerner, A. P.: "The Concept of Monopoly," 23n
Levine, A. L.: *Industrial Retardation in Britain*, 3n; generalizes from iron and steel, 2
Lindert, P., and Trace, K.: "Yardsticks for Victorian Entrepreneurs," 12n
Lincolnshire (East Midlands) ores. *See* Basic process; Chapter 4, *passim*
Lloyd's Register: "Extended Report on Steel," 48n
Location of British iron and steel: scattered over ore and coal fields, 23; interregional competition, 29–33; Northeast Coast (Cleveland) center of pig iron, 32, 144; no fragmented local markets, 37; importance for testing effect of demand, 41; Scotland and Northeast Coast centers of shipbuilding steel, 54n; East Midlands as location for pig iron, Chapter 4, *passim;* stable source of ore implies constant richness, 79

McCloskey, D. N.: ed., *Essays on a Mature Economy,* 11n; "International Differences in Productivity," 12n; "Did Victorian Britain Fail?" 13n. *See also* McCloskey, D. N., and Sandberg, L.
McCloskey, D. N., and Sandberg,

L.: "From Damnation to Redemption," 12n
Macrosty, H. W.: *The Trust Movement,* 26n
Managers. *See* Entrepreneurial failure in Britain, historiography
Manufactures, Census of, U.S., 100
Market scale of British iron and steel: measured by value of gross output, 19, 140–142; growth in ship plates relative to rails explains strong competition in plates, 34; slow growth no obstacle to large plants, 36–37; measured by output of pig iron, 38n; relevance to issue of failure, 40–42; growth dependent on domestic investment, 41–44; of steel, dependent on substitution of steel for iron, 44–45; growth incorrectly used as measure of success, 4–5, 6, 39–40, 44–45, 55, 127. *See also* Embodiment; Exports; Railmaking steel; Shipbuilding steel; Chapter 3, *passim*
Market structure. *See* Monopoly in British iron and steel
Marshall, A.: "Fiscal Policy of International Trade," 6n; *Principles of Economics,* 6n; *Industry and Trade,* 6n; Britain loses leadership, 6
Mathias, P.: *First Industrial Nation,* 8n
Meyer, J. R. *See* Conrad A. H., and Meyer, J. R.
Minhas, B. S. *See* Arrow, K. J., and others
Mitchell, B. R., with Deane, P.: *Abstract of British Historical Statistics,* 16n
Monopoly in British iron and steel: narrative inadequate for measurement, 22–23; entry and large numbers discourage collusion, 22–23, 27–28, 33–34; Lerner's

Index

Monopoly (*cont.*)

measure introduced, 23–24; small by cycles in prices of outputs and pig iron, 24–28; railmaking monopolized after *1895,* 25–28, 32, 34; absent in shipbuilding, 25, 28, 32, 34; effects of capacity utilization on measure, 25n; rigid prices confirm measure, 26; small by prices at home and abroad, 29–33; causes in rails and ship plates compared, 34

implications of absence in iron and steel: for family firms, 35–36; for economies of scale, 36–37; for managerial skill, 37–38

effects on measures of productivity, 22, 82n

Mundlak, Y.: "Empirical Production Function," 77n

Mushet, R.: spiegeleisen, 48

National Federation of (British) Iron and Steel Manufacturers: *Statistics,* 40n

Nelson, R. R.: "Aggregate Production Functions," 108n

Open hearth process: early improvements, 47–48; superior for shipbuilding, 48–50, 54; fall in cost causes adoption, 50–53; most common in U.K. after *1880*'s, 53–54, 141; shipbuilding takes most output, 54; little scrap used in U.K., 72; adoption of basic process depends on Talbot furnace, 70–72. *See also* Productivity, rate of change, ship plates; Shipbuilding steel

Ore, iron: North Yorkshire (Cleveland) compared with Lincolnshire (East Midlands), Chapter 4, *passim;* effect of quality on productivity in pig iron, 78–79, 115–116

Output. *See* Exports; Market scale of British iron and steel; Railmaking steel; Shipbuilding steel

Payne, P. L.: "Iron and steel Manufactures," 21n; difficult to measure entrepreneurship, 20–21

Phelps-Brown, E. H., and Handfield-Jones, S. J.: "The Climacteric of the 1890's," 7n. *See also* Climacteric in British productivity

Phelps-Brown, E. H., and Weber, B.: "Accumulation, Productivity and Distribution," 76n

Pollard, S.: "Wages and Earnings in the Sheffield Trades," 91n

Pollard, S., and Crossley, D. W.: *The Wealth of Britain,* 8n

Pools. *See* Monopoly in British iron and steel

Potter, E. C.: "The South Chicago Works," 118n

Prices of iron and steel: abundance and usefulness as evidence emphasized, 24, 85, 87, 95, 98–99, 120; tables, 134–139

Production, First Census of (U.K., *1907*), 16n

Productivity

theory: gross costs relevant, not value added, 17–19, 74, 92; price measure of productivity change, 25; 34, 85–86; biased toward no difference in levels when activities in same market, 66; shares in costs correct weights for input, 75; share of entrepreneurship, 76–77; input with largest share best for partial measure, 78; substitution effect in partial measures, 80, 116–119; shares

154

Index

in value correct weights for
outputs, 95–96; Jones's
"real costs" identical to
price measure, 103
sources of mismeasurement:
monopoly, 82n; economies
of scale and nonneutrality,
82–83; varying utilization of
capacity, 82, 124, 147;
errors in inputs, 83–84, 88–
89, 144, 145
embodiment effect, 106–107,
109–111
solution of index number
problem, 143
rate of change: relevance to
entrepreneurial failure, 12–
13, 73–74, 76–77; rapid pace
contrains monopoly in ship
plates, 34
measured in entire economies:
United States, 14; Germany,
14; data for U.K., 131–133
measured in iron and steel:
U.K. pig iron, 77–84; U.K.
all iron and steel, 92; U.S.
rolled steel, 95–101; Belgian
railmaking, 99; U.S.
Bessemer and open heart
separately, 99–101; U.S. pig
iron, 101–104; U.S. all iron
and steel, 104. See also
Chapters 5 and 6, passim
differences in levels in iron and
steel: imply opportunities for
profit, 12, 77, 125; relevance
to entrepreneurial failure,
21; no opportunities for
profit in faster adoption of
basic steel and basic ores,
Chapter 4, passim; U.S. as
a standard of perfection,
94–95; size between U.S.
and U.K. implied in literature, 105, 114
size implied by rates of change

given initial equality: in
steel, 101, 105–112, 126; in
pig iron, 103, 126n
measured: in steel, 120–124;
in pig iron, 114–119
See also Embodiment

Railmaking steel: monopolies in,
25–28, 32, 34; productivity in
U.K., 34, 84–90; adoption, 46;
productivity in U.S., 99–101;
productivity in Belgium, 99;
levels of productivity compared
in U.S. and U.K., 120–124. See
also Bessemer process
Railways: chief example of interrelatedness, 9–10; unimportance
of entrepreneurial failure in, 20n;
See also Transport costs
Reuss, C., and others: Le Progrès
Economique en Sidérurgie, 99n
Report of the Tariff Commission,
64n
Rosovsky, H., ed.: Industrialization
in Two Systems, 41n
Royal Commission on Canals and
Waterways, 63n

Samuelson, B.: "Construction and
Cost of Blast Furnaces," 118n
Sandberg, L.: "American Rings and
English Mules," 12n. See also
McCloskey, D. N., and
Sandberg, L.
Saul, S. B.: Myth of the Great
Depression, 8n; narrative studies
of engineering contradict
hypothesis of failure, 8
Sayers, R. S.: History of Economic
Change in England, 7n
Schumpeter, J. A.: entrepreneur in
economic history, 6–7
Shephard, R. W.: Cost and
Production Functions, 86n
Shipbuilding steel: competitive
British market in, 25, 28, 32–33,

155

Wright, C. D.: *Sixth Annual Report of the Commissioner of Labor,* 104n
Wrought (puddled) iron: highly competitive, 25, 33; steel substitutes, 44–53; scarcity rents of puddlers dissipated in *1880*'s, 88; productivity change probably nil, 92; small share of output by *1907,* 140–141

Yasuba, Y.: "Profitability and Viability of Plantation Slavery," 62n